Local Economic and Employment Development (LEED)

Clusters, Innovation and Entrepreneurship

OECD

ORGANISATION FOR ECONOMIC CO-OPERATION AND DEVELOPMENT

The OECD is a unique forum where the governments of 30 democracies work together to address the economic, social and environmental challenges of globalisation. The OECD is also at the forefront of efforts to understand and to help governments respond to new developments and concerns, such as corporate governance, the information economy and the challenges of an ageing population. The Organisation provides a setting where governments can compare policy experiences, seek answers to common problems, identify good practice and work to co-ordinate domestic and international policies.

The OECD member countries are: Australia, Austria, Belgium, Canada, the Czech Republic, Denmark, Finland, France, Germany, Greece, Hungary, Iceland, Ireland, Italy, Japan, Korea, Luxembourg, Mexico, the Netherlands, New Zealand, Norway, Poland, Portugal, the Slovak Republic, Spain, Sweden, Switzerland, Turkey, the United Kingdom and the United States. The Commission of the European Communities takes part in the work of the OECD.

OECD Publishing disseminates widely the results of the Organisation's statistics gathering and research on economic, social and environmental issues, as well as the conventions, guidelines and standards agreed by its members.

Also available in French under the title:
Développement économique et création d'emplois locaux (LEED)
Pôles de compétitivité, innovation et entrepreneuriat

Corrigenda to OECD publications may be found on line at: *www.oecd.org/publishing/corrigenda*.

Foreword

Provision of policy guidance on the development of local clusters of entrepreneurship and innovation is one of the key work areas of the OECD Centre for Entrepreneurship, SMEs and Local Development. Clusters are one of the most remarkable features of local economic development and small and medium-sized firm competitive advantage. Localised groupings of suppliers, customers, competitors and support institutions, all specialised in related activities, are present in all economies. They are particularly productive locations for the economic activities they host. They are sites of strong entrepreneurship and innovation vitality. And, because of this, they drive economic and employment growth. But what makes clusters grow? What threats do they face? How should policy intervene? Answering these questions is the key motivation of this book.

The governments the OECD works with at national and local levels, and the development agencies they support, are all aware of the concept of clusters and the potential that clusters hold for stimulating economic development and many have put in place their own policy initiatives for clusters. This has led to strong demands on the OECD for information on how policies can strengthen clusters based on an international assessment of recent experiences.

Here we share information from this work on success factors, problems and policy approaches in major clusters. The book shows the importance to cluster growth of co-operation, research commercialisation, human capital and skills enhancement, public sector commitment, partnerships and leadership, quality of life and social capital. It also shows the challenges that even strong clusters face. Policy is already doing much to support cluster development in the places reviewed in this book. A number of recommendations are presented for other areas based on the assessment of this experience. Key messages include the importance of encouraging new firm creation, stimulating innovation, co-ordinating polices, ensuring human capital, facilitating access to finance and addressing congestion and social divisions.

The clusters examined in this book all experienced a long period of growth in the decade to 2008, but since the past few months are confronted with the major challenge of responding to a global economic crisis. The crisis has turned attention towards survival, restructuring and downsizing in the face of shrinking markets and restricted finance. In this new context, the favourable environments that clusters provide to the enterprises they host remain extremely important as fundamental sources of competitiveness. Indeed, the productivity and cost advantages that clusters offer should help their firms to weather the crisis better than firms that stand alone. But decisions and investments made in clusters now also affect their prospects for growth and employment creation in the upturn. This is therefore a time for smart policy, which identifies the critical investments and actions that need to be taken to reinforce cluster strengths and position them as sources of future entrepreneurship, innovation and growth. The ideas in this book show the directions that need to be taken.

It is the result of work undertaken by the Local Economic and Employment Development Committee of the OECD in collaboration with the French Assembly of Chambers of Commerce and Industry, the Caisse des Dépôts et des Consignations, and the French Ministry of Economy, Industry and Employment, who wished to compare the experience of Grenoble with that of clusters in other countries. I hope it will stimulate the development of effective cluster policies and practices in a wide range of settings.

Sergio Arzeni
Director, OECD Centre for Entrepreneurship,
SMEs and Local Development

Table of Contents

Box

Tables

CLUSTERS, INNOVATION AND ENTREPRENEURSHIP – ISBN 978-92-64-04442-5 – © OECD 2009

Figures

Acknowledgements

The OECD Local Economic and Employment Development (LEED) Programme would like to acknowledge the financial contributions of the French Ministry of Economy, Industry and Employment (Directorate for Commerce, Handcrafts, Services and Professionals and Directorate General for Enterprises [DCASPL]), the French public investment bank (the Caisse des Dépôts et des Consignations) and the French national assembly of chambers of commerce and industry to this publication.

In addition, the chapter on the Grenoble cluster benefitted from advice and substantive contributions from the DCASPL and the National Statistics Department (INSEE). Various French institutions were actively involved in the steering group, which met twice to discuss the draft and conclusions of a report on the Grenoble cluster in comparison with other experiences which forms the basis for this publication. Representatives of the Délégation interministérielle à l'aménagement et à la compétitivité des territoires (DIACT), the Délégation générale du pôle Minalogic, the Préfecture de la région Rhône-Alpes, the Conseil régional Rhône-Alpes, and the Conseil général d'Isère provided useful comments.

The policy recommendations contained in this publication have been endorsed by the LEED Directing Committee at its 51st session held in Paris.

Executive Summary

Contrary to popular belief, the globalised knowledge economy relies more and more on the local dimension. In order to boost economic development and respond efficiently to ever keener international competition, OECD member countries need to pursue entrepreneurship and innovation policies that reflect their own distinctive local characteristics. Throughout the OECD area, innovation is increasingly concentrated within clusters of enterprises and research/training institutions that work on complementary activities. Indeed, it has been demonstrated that clusters are an important source of innovation and competitiveness driven at the local level. Clusters create an environment conducive to productivity gains, which are a factor of growth, and so form a structure that helps enterprises meet the challenges of international competition. This local dimension of innovation and entrepreneurship nonetheless poses challenges to policy makers because clusters require policies and support schemes that are tailored to local needs.

According to Porter, clusters are geographic concentrations of interconnected companies, specialised suppliers, service providers, firms in related industries, and associated institutions in particular fields that compete but also co-operate. This book examines policy approaches and policy recommendations for cluster development based on the in-depth analysis of seven internationally reputed clusters in the OECD area, namely Grenoble in France, Vienna in Austria, Waterloo in Canada, Dunedin in New Zealand, Medicon Valley in Scandinavia, Oxfordshire in the United Kingdom, and Madison, Wisconsin, in the United States.

What are the key factors for cluster success? What is the appropriate level and type of public intervention for cluster development? The book addresses these and other issues based on the analysis of the seven cluster studies. It looks at the factors that have contributed to the successful development of each cluster and at the barriers that hinder their expansion. Following this analysis, the review puts forward a set of policy recommendations that are geared to the context of clusters.

Main findings

The analysis shows how clusters contribute to the dynamic for enterprise creation, notably by stimulating the creation of highly innovative spin-outs in core cluster sectors. Moreover, business creation in the economy in general seems also to be positively influenced by clusters through increased demand for indirect services (*i.e.* induced activities). For instance, in 2007, 40 per cent of the firms in the heart of the Grenoble cluster had started up since the labelling of the "pôle de compétitivité" by the government in 2005. Furthermore, among the SMEs of this grouping, more than one in three had started up in 2006.

The book also shows how clusters can have a positive impact on employment, both directly, in the high-tech enterprises that form the heart of the clusters, and indirectly, through their ripple effect in the economy. In Oxfordshire, for instance, from a base of 190 enterprises with 2 000 employees in 1989, the region counted approximately 3 500 high-technology enterprises with 45 000 employees by 2004. In the Madison cluster, in the United States, employment increased by 75 per cent between 1993 and 2004 in the core sectors of the cluster, life sciences and information technologies.

Key factors for success

Strong co-operation. In all of the clusters examined, there was a strong fabric of enterprises, reputed universities and top level research centres that join forces to coproduce innovation. Strong co-operation is a key factor in successful innovation and, more specifically, collaborations between research organisations, universities and enterprises in the form of start-ups and SMEs. Such co-operation has often been fostered by agencies or networks outside the cluster acting as brokers. The interaction between players in a cluster may grow with the development of a policy to promote local networks and supply chain development. The importance of such an approach is clear in Medicon Valley, where government policy has sought to build up the networking capacity of players in the "triple helix" (government, industry and research).

Research commercialisation. It is crucial for successful spin-off activity to identify the market for an innovation. However, this is not always simple and its importance is often underestimated. Very few efforts have been made to teach students or advise researchers on how to commercialise their innovative product to make a profit. Furthermore, the identification of the market is a key step in attracting private investors in high-tech sectors and ensuring the investment to build clusters, notably venture capital. The WAVE programme in the Wisconsin cluster is an initiative that has proven to be successful in teaching students and assisting academics commercialise their innovations.

Critical human capital mass. A pool of talents is the base for the emergence and development of clusters. Hence, the education supply and cluster demands should be matched in order to avoid a shortage of skilled labour and therefore a flight of financial capital and know-how. The availability of qualified people at various levels needs to be forecasted and guaranteed so that the cluster can evolve at the same pace as innovations. A special effort is needed to attract highly qualified labour. Some initiatives to attract talents from abroad have worked successfully in the Vienna cluster (targeting expatriates) and in the Dunedin cluster (targeting foreigners).

Skills enhancement. The adaptability of clusters to technological and competitive change must not be undermined. Local players are engaged in international competition and must know how – and be able – to react to developments in their markets. High skill levels help cluster enterprises to adapt to changing conditions. Furthermore, it is important to adapt training provision to new areas of demand in clusters. One particular aspiration is the provision of training for emerging activities not only in engineering but also in supporting technical professions. Many international clusters have set-up training programmes to cover emerging fields, as is the case of Waterloo and its Co-op programme.

Strong commitment of the public sector. Public actors have played an important role in the emergence and development of the clusters examined in this book, regardless of the cluster's nature. For instance, the public sector has apparently played a limited role in the development of the Madison cluster (United States). Yet in Madison there is a comprehensive strategy and in the US there is substantial public funding at the national level for academic research and numerous government funding programmes for technology spin-outs that the cluster has benefited from. If clusters are to remain competitive in the long-term, the public authorities should help promote a comprehensive strategy and agree on investments in infrastructure and public services such as transport, housing, schools and personal services to meet new cluster requirements.

Strong partnerships and leadership. For all of the clusters analysed, strong partnerships and clear leadership have been essential factors for success. Public-public partnerships are the key to ensuring a co-ordinated overall policy strategy for the development of the cluster. Public-private partnerships are crucial in making sure that policies are adequate and that needs are met in the cluster. For instance, in the case of Grenoble, the involvement of local authorities in the governance of the Minalogic cluster management organisation has helped to forge relationships based on trust with its members. From the outset, strong partnerships and the leading role played by the cluster's delegate-general has contributed to the success of Minalogic. Moreover, partnerships among firms and between research and industry are critical for supporting collaborative research and training initiatives and knowledge transfers within clusters.

High quality of life. A high quality of life has also proved to be an important factor in retaining, renewing and growing the critical mass of talents in the clusters reviewed here. Social networks, people-oriented services, low bureaucratic procedures and good infrastructure are just some of the factors that have an impact in the attraction and retention of highly skilled population into a given place. The Grow Wisconsin Initiative in the Madison cluster is one of the approaches recently adopted by the regional government to guarantee a high quality of life and an environment attractive not only to talents but also to new enterprises. Other similar initiatives such as the Smart Growth initiative in the Oxfordshire cluster are important efforts made to ensure a good quality of life.

Social capital. Finally, it is important to stress that social capital is a major intangible asset for all of the clusters reviewed that has certainly contributed to their success. Innovation cannot take place in a hostile environment, and it is therefore important to encourage the creation and strengthening of networks of trust and exchange. For instance, in Grenoble, the "co-ompetition" among the members of the cluster, as well as the ongoing dialogue between public and private players, make the city a particularly attractive place for innovation. Grenoble is an illustration of how a society based on trust and co-operation can help to develop human potential. In Grenoble, innovation goes hand in hand with trust.

Barriers hindering cluster development

Weak entrepreneurial cultures and incentives in research. Although there is significant spin-off activity in the reviewed clusters, their development could be stronger if barriers were overcome to business creation from research centres and universities and in the impact of often rigid regulations on intellectual property rights. Prevalent research and academic cultures are not good at stimulating researchers to set-up firms. Furthermore, those ready to make the move are often hindered by the legal frameworks of university systems. Intellectual property bureaus at universities could play a role in changing this problem. Initiatives such as the "Best of Biotech" contest in the Vienna cluster which seeks to encourage academics to start their own business could also have a positive impact in business creation.

Weak involvement of small firms in cluster projects. Clusters do not usually integrate small enterprises and firms in traditional sectors into their core research, development and innovation activities. In some cases, these enterprises are seen to lie outside the sphere of policy support and collaborative projects. However, innovation in these firms could be boosted by closer participation in cluster activities. Initiatives such as the Metis project in Grenoble, involving closer co-operation between textile companies and high-tech firms, can be beneficial.

Lack of seed capital. A common barrier to entrepreneurship development is lack of seed capital for the start-up and growth phases of firm development.

Although this problem may be attenuated in clusters compared with other locations, there is a need for public support to help overcome these problems even in the reviewed clusters.

Shortages of qualified labour. Skill shortages were present in some of the clusters extending from the very highly skilled to medium and high skill support workers. The growth of clusters and the rapid evolution of their skills needs present education and training providers with difficulties in delivering the right skilled graduates to cluster enterprises. It is often useful for education and training providers to work with existing firms in upgrading and adapting their workforces to new needs.

Problems of congestion and social divisions. The development of clusters may create or aggravate problems of social cohesion as it increases the demand for highly skilled labour but has only a small direct impact on unskilled or low-skilled workers. Clusters are also experiencing congestion in housing, land and transport, creating problems for existing residents. Furthermore, there are examples of resistance to research activities undertaken in areas where ethical concerns are important. As a consequence, it is common to find lobbies opposed to clusters and cluster growth. This issue should not be viewed as trivial, but should be taken seriously and dealt with rapidly to prevent such movements from spreading and undermining the cluster's expansion. Good communication helps to address this problem. The Oxfordshire cluster, for instance, produces regularly a newsletter on the activities of the cluster to inform the non-scientific community on ongoing and future projects.

Poor co-ordination of policies. There is usually a large body of local actors involved in various aspects of cluster support, and often they are not well co-ordinated. In addition to these local actors there are often national actors from various ministries with various programmes and agendas. If these actors could be pulled together more effectively, policy could better target and respond to the key needs of the clusters. The development of local cluster strategies would help to respond to this problem.

Policy recommendations

The analysis of the reviewed clusters provides a basis for a set of recommendations on the emergence and development of clusters and entrepreneurship elsewhere. These recommendations are listed below.

1. *Encouraging entrepreneurship*

For clusters to survive in the long-term, it is crucial to actively foster entrepreneurship in order to promote the creation and growth of start-ups that can contribute to the cluster's development as suppliers, partners or clients. The enterprise fabric of clusters needs to be supplied and renewed with highly innovative enterprises issued from the universities, research centres or other large companies. This can be achieved by:

- *Supporting spin-outs and small firm collaborations*: by encouraging spin-outs and collaborations among large enterprises, research institutions and SMEs, especially through relations with suppliers and subcontractors.
- *Leading a transition to the entrepreneurial university*: by promoting entrepreneurship at all levels of education and in all fields, emphasising issues related to the capacity to create and market innovations.
- *Supporting the launch and growth of start-ups*: by promoting a culture of entrepreneurship in the clusters, implementing specific programmes to finance enterprising projects, and fostering the emergence of skills pools to support and work with these new enterprises.

2. *Stimulating innovation and collaboration*

It is crucial to consolidate dialogue between universities, industry and laboratories to provide better chances of collaboration and therefore enhance innovation. Also, the promotion of firm networks can contribute to the creation of a climate of trust between the members of the cluster and strengthening supply chains. Innovation and collaboration can be stimulated by:

- *Fostering industry – research collaboration*: by promoting the development and transfer of technologies within clusters through technology transfer centres, neutral agencies or networks serving as brokers, and encouraging mobility among professionals between industry and the academic world.
- *Encouraging enterprise networks*: by introducing SMEs into formal networks or by providing regular platforms for social meetings to generate a "cafeteria" effect.
- *Stimulating spin-offs*: by creating policies to increase the ownership by "creators" in the university of intellectual property rights over the results of research.
- *Better marketing of products*: by the establishment of marketing centres for products resulting from academic research located in universities.

3. *Co-ordinating public polices and local initiatives*

Co-ordination at all levels of government is essential to ensure the relevance of policy design, the integration of policies into an overall strategy and the achievement of efficiency in policy delivery. From the creation to the development of clusters, partnerships have a major role to play as facilitators of policy coordination. The co-ordination of public policies and regional initiatives can be supported by:

- *Strengthening public-public and public-private partnerships*: by bringing together key enterprises, local and national government authorities, university institutions and the business community to work together on the definition of common initiatives.
- *Encouraging evolution in cluster activities*: by identifying activities in complementary knowledge sectors to the existing base of the cluster and providing new infrastructure, new networks and the development of new enterprise formation.

4. *Ensuring quality human capital*

Clusters ought to attract human capital in order to attract, retain and embed financial capital (private investors, foreign direct investment, etc.) and produce innovation. The best place to live has become the best place to invest and work. It is therefore crucial to maintain a high quality of life in order to guarantee the availability of highly skilled talents in clusters. This can be achieved by:

- *Updating education and training to meet the requirements of the cluster*: by establishing a continual dialogue between industry and universities so as better to understand those needs, and by creating trend databases and forecasts of skills needs to avoid skills shortages.
- *Ensuring availability of talent locally*: by attracting talents from abroad to come to work in the cluster through the implementation of programmes which target foreigners or expatriates.
- *Ensuring the appeal of the area and a good quality of life*: by tackling problems affecting the quality of life in a region, such as road traffic, high housing prices, scarcity of international schools, poor quality public services, etc.

5. *Facilitating access to financing*

Clusters face global competition and any delays in access to finance can be critical to the development of the cluster and its position in the global market. As most of the innovations in high-tech sectors require important amounts of funds with long periods of incubation, private investors often lack incentives to invest in a cluster. It is therefore necessary to facilitate a rapid access to public funds and to encourage private investment. This can be achieved by:

- *Encouraging private investment*: through a technology transfer bureau, or with specific tax incentives for funds placed in an innovation project in the cluster.

- *Facilitating access to public funding*: by reducing bureaucratic procedures and by creating and publicising a "one stop shop" for SMEs.
- *Creating forums to seek financing*: by organising recurrent well-known events which serve as platforms for investors and entrepreneurs to meet.

6. Reducing congestion and social divisions

Clusters need to operate transparently and have an inclusive approach to the community in order to avoid social divisions. Many tensions can emerge in the region from the establishment of a cluster, which should be dealt with as soon as they are perceived in order to remove barriers to the development of the cluster. This can be achieved by:

- *Tackling congestion and social inequalities resulting from the emergence of the cluster*: by offering training programmes on the new skills required in the cluster to the unemployed and by investing in housing, infrastructure and public transport following a strategic plan.
- *Creating mechanisms to inform about the activities of the cluster*: by issuing periodical newsletters addressed not only to the scientific community but also to the population and other enterprises not directly involved in the core activities. This could facilitate the identification of new business opportunities in the community (induced activities) and give a positive image of the cluster.

CLUSTERS, INNOVATION AND ENTREPRENEURSHIP – ISBN 978-92-64-04442-5 – © OECD 2009

ISBN 978-92-64-04442-5
Clusters, Innovation and Entrepreneurship

Chapter 1

Policy Issues in Clusters, Innovation and Entrepreneurship

by
Jonathan Potter, Centre for Entrepreneurship,
SMEs and Local Development, OECD

This chapter sets out the key policy issues examined in the book on supporting entrepreneurship and innovation through clusters. It discusses the nature and importance of clusters, how clusters function, and the role of policy. Clusters stimulate entrepreneurship and innovation because they are sites of localised positive externalities in labour market pooling, input-output linkages and knowledge spillovers. Policy has a role to play in facilitating the emergence and growth of clusters and addressing some potential problems of clustering.

Introduction

Major localised clusters of firms and research and training institutions in closely related activities play an important role in increasing innovation, entrepreneurship and growth in the local economies that host them, generating both local and national economic benefits. This book looks at how selected successful clusters operate, how they are supported by policy and the lessons for other areas. We use a local lens to understand these clusters and cluster policies because each cluster is situated in particular local conditions and the relevant policy interventions must therefore be tailored to their distinct needs.

Clusters support economic growth by providing an environment conducive to innovation and entrepreneurship in given activities, which enables productivity growth through increases in efficiency and the quality and differentiation of goods and services. Many of the benefits occur naturally, without any need for public intervention: knowledge spillovers happen when workers change firms and entrepreneurs spin-off ideas from large firms or research institutes; labour search and training costs are reduced when there are large local pools of specialised labour; and firms capture internal economies of scale when selling specialised inputs to large local markets. But public policy can also play an important role in the emergence and development of successful clusters, for example by supporting pre-competitive research collaborations, providing basic and specialised infrastructure, brokering network creation, funding generic education and training and helping to co-ordinate and mobilise key stakeholders behind a vision for the development of the cluster.

This book provides seven case studies of the recent evolution of major, successful clusters in OECD member countries, the challenges they face and the role that policy is playing there. Each cluster is crucial to its local economy and of real significance to its national economy. Their development is complex and context specific, but examining their stories raises important issues for cluster development elsewhere. The story of the Grenoble cluster in France is illustrative.

The economy of the Grenoble city-region has grown strongly in recent years, driven by an internationally competitive cluster of activities in research, development and product design for microelectronics, nanotechnologies and related software. It counted in 2007 some 40 000 direct jobs in 300 enterprises,

CLUSTERS, INNOVATION AND ENTREPRENEURSHIP – ISBN 978-92-64-04442-5 – © OECD 2009

of which the majority were less than 10 years old, and included many spin-off enterprises. Cluster output and employment has grown strongly in the last few years, especially within SMEs, and its firms undertake a high level of innovation, with over 80 per cent bringing new products to market in a 3-year period. In 2005, the cluster was recognised by the French government as one of seven "global competitiveness clusters" in France, or "pôles de compétitivité mondial", bringing considerable financial support for research and development activities and a cluster management organisation, Minalogic, with the job of brokering research and other collaborations between research, education and industry.

The development of the Grenoble cluster has nonetheless unfolded over a long period of time. The development of electronics and microelectronics activities in the area can be traced back for at least 50 years to the development of hydroelectric power in the surrounding Alps and the location of a branch of the government's atomic energy agency undertaking industrially relevant research in the city. Following these largely informal and fortuitous origins, it is interesting how the cluster then gathered strength and grew over time, successfully updating its technologies by combining and recombining a number of technology staples, from the era of hydro-electric power, electrochemistry and electrometallurgy in the 1950s to the present era of nanotechnology.

What has been behind this success? Views differ on the details and precise order of importance, but certain elements stick out. Grenoble has built on a very strong education sector, comprising four universities and a management school delivering a large inflow of high-skilled labour in relevant fields. It has eight major national or international research institutes in related fields. Strong local social capital at leader level and an entrepreneurial and pro-cluster local public sector has helped to drive research-industry co-operation and public-private investment projects. Some major firms have located their research and development facilities locally, which despite subsequent changes of ownership have remained locally embedded and strong sources of spin-off enterprises. And there have been some important flagship projects in the last ten years, including the Alliance Crolles 2 initiative, an agreement amongst three major firms to pool some of their pre-competitive research efforts, the Minatec project, a centre for joint education, training and research for the cluster, and the recent "pôle de compétitivité" initiative, each driving the cluster forward. So in a strong sense the growth of the Grenoble cluster has been supported and accompanied by cluster-oriented public sector activity and investment and strong local leadership.

But there are new challenges for the cluster, which potentially threaten its continued growth. One major issue is the need to respond to intense and increased international competition. Major investments in platform technologies were made in the past, but a new scale of investment is now implied for a new generation of microelectronics products, and the cluster is seeking to compete with others abroad that are benefiting from substantial public investments. A second challenge is to reverse a worrying, if small, recent decline in the number of researchers in the cluster and a shortage of technicians. And there is a continued "invisible barrier" between industry and public research and education institutions that constrains the labour mobility that promotes knowledge transfer and collaborations. SMEs have also been kept too far away from formal cluster governance arrangements, limiting their potential role in contributing to and disseminating innovations and spreading job creation benefits through subcontracting. Then there is a major question concerning the future trajectory of the cluster, with the key industry, research and government players needing to strike the right balance between specialisation and diversity in the cluster, seeking on the one hand to minimise the risk of becoming a single specialisation cluster vulnerable to shocks and on the other to avoid a fragmentation of research and training efforts that would reduce critical mass. This challenge is epitomised by a recent local debate about whether public research funding for the cluster should move into new areas such as sophisticated textiles that share a common competence and skills base with the traditional cluster specialisation but are not strictly in the fields for which the cluster is best known. Furthermore, success itself has generated new challenges and threats in terms of managing congestion in transport and housing, limiting the risk of social divide by ensuring that growth benefits are spread widely in the city and not limited to an elite group of high-tech employees and responding to ethical objections to nanotechnology research in the cluster, for example by sharing information and developing ethics policies.

There are clearly some potentially important lessons from this story for other clusters. The case of Grenoble demonstrates the potential of a local cluster to drive growth through generating and supporting innovative enterprises able to compete on a world stage. It underlines the critical role that public policy often plays in supporting the development of clusters. It shows the role that leadership, social capital, public investment and collaborations have played in supporting development. It also shows that clusters need to evolve over time through adaptation to changes in technologies and markets. And it suggests that clusters should not be seen as assured successes, able to look after themselves and spread benefits more widely, but that their growth and evolution are likely to bring with them new challenges often requiring further co-ordination amongst stakeholders and targeted policy support.

Cluster policies are still relatively new in many places and substantial policy experimentation is taking place. There is much to learn from these experiments about the role of policy and its methods and approaches. This book takes up this

CLUSTERS, INNOVATION AND ENTREPRENEURSHIP – ISBN 978-92-64-04442-5 – © OECD 2009

learning challenge by seeking answers to the following questions for each of seven internationally important clusters (Grenoble, France; Vienna, Austria; Waterloo, Canada; Madison, Wisconsin, United States; Dunedin, New Zealand; Oxfordshire, United Kingdom; Medicon Valley, Sweden/Denmark):

- What is the nature of the cluster and how has it evolved over the last ten years?
- What are the key drivers of competitiveness of the cluster and the factors behind its success?
- What is the impact of the cluster on entrepreneurship and new and small firm development?
- What are the obstacles to further development of the cluster?
- What is the role of public policy at national, regional and local levels in supporting the cluster?
- What are the future policy challenges for the cluster?
- What are the lessons of the experience for clusters elsewhere?

By reviewing these experiences, the book aims to point to the most promising directions for policy development in other clusters.

But first this introduction addresses some key preliminary issues that should be understood before deciding on policies for clusters, namely: What are clusters? Why focus on clusters? What are the limits to cluster approaches? How do clusters function? What is the role of policy in clusters?

What are clusters?

Clusters are geographic concentrations of firms and organisations working in related activities. They have been defined by Porter (1998) as geographic concentrations of interconnected companies, specialised suppliers, service providers, firms in related industries and associated institutions in particular fields that compete but also co-operate. They are often rooted in a single locality, but their boundaries will not always correspond to local administrative areas, as illustrated by the Medicon Valley case in this book. They exist across the full array of tradable sectors although they often do not correspond to a single manufacturing or service sector as recorded by standard industrial classifications.

The feature of tradability of cluster output across local and national boundaries is an important one. First, it highlights the fact that clusters of locally-concentrated activity can only emerge when it is possible to trade their output, since otherwise the scale of the activity is limited by the local population which it serves. Second it highlights how the size and specialisation of clusters have been able to grow given globalisation and increased international market integration. The potential impacts of clusters are therefore much greater now than in an era of stronger national trade barriers and higher transport costs. Third, because they are tradable and grow into international markets clusters are of course subject to strong competition from other places.

From a national or international perspective, the economic benefit of clusters comes from the specialisation of places in activities within which firms can gain higher productivity through accessing external economies of scale or other comparative advantages and the trading of this higher quality and lower cost output with other places. From the perspective of local governments, the benefit comes from the strong competitiveness of the clusters they host, their generation of higher wages and profits, their attraction of new investment and the spread of benefits to the rest of the local economy.

Clusters may be supported by a specific cluster initiative or programme, for example a network broker or manager, but this is not always the case. Indeed, a frequent mistake made by policy makers and analysts is to think that clusters are synonymous with deliberate policies or deliberate co-operations in formal networks. Although this type of action is common and can be important, clusters may also arise and sustain themselves simply through market processes as new actors move into a cluster to gain the benefits of greater productivity and existing actors grow through their greater competitiveness. Thus the role of policy has to be carefully assessed.

It is also important to recognise that there are different types of clusters with different methods of functioning. Iammarino and McCann (2009) identify three types of clusters: "pure agglomerations" such as a competitive urban economy with fragmented and hard-to-identify relations among firms, "industrial complexes" such as steel dominated clusters with strong vertical linkages in local supply chains, and "social networks" such as new industrial districts with strong social capital based relationships. A classic article by Markusen (1996) distinguished between "new industrial districts", "hub and spoke districts", "satellite industrial platforms", and "state-centered districts". Panniccia (2006) provides another useful typology. The point is that the nature of clusters and the processes that make them competitive differ, and these differences must be understood if appropriate policy responses are to be developed.

Why focus on clusters?

Economic benefits

For those involved in policy development, the fundamental motivation to understand clusters is to explore what policy might do to facilitate their emergence and growth and hence realise the potential economic benefits associated with them. The picture is clearest when the lens is placed on those places that host clusters. Within these places cluster development may be associated with productivity increases resulting from a range of localised external economic benefits discussed further below. These productivity increases allow increases in the wages and profits or price competitiveness of local enterprises. Furthermore, they can create a virtuous circle of growth as the higher productivity environment generates new firm entrants and firm expansions, increasing the size of the cluster and strengthening the productivity benefits. Local growth may therefore be expected in terms of income per head increases and job creation.

From the national perspective, the picture is more complex because the external economic benefits in the locations where clustering occurs may be accompanied by external economic costs from losses of activity in other locations. If the productivity gains within the cluster outweigh the productivity losses in other places, then national rather than just local gains can be expected. Whether such net national gains exist is a difficult empirical question to answer. Nonetheless, theory suggests that if a range of cluster specialisations emerge across a country then national benefits might be expected from the increased division of labour. In addition, major clusters compete internationally and as such may generate benefits without necessarily displacing other national companies from their product markets. But critically, because of increasing returns more innovation may be generated from location in clusters, thus stimulating growth. Furthermore, the growth of major clusters may also spread more widely within a country through mechanisms including purchasing from suppliers outside of the cluster, transfer of knowledge to other areas, and redistribution of tax revenues.

Political interest

Growing understanding of the potential for realising these types of economic benefits has also raised political interest in clusters in recent years. The French competitiveness clusters ("pôles de compétitivité") programme has already been highlighted in the case of the Grenoble cluster. It is an initiative that aims to promote innovation, exports and growth in French strategic industry sectors by building critical mass in 71 clusters, with a greater weight of spending going towards the 7 global competitiveness clusters. Important national cluster initiatives can also be found in other OECD countries, such as the Technology Cluster Initiatives in Canada, the Czech Clusters Programme and the Japanese Industrial Cluster Programme (OECD, 2005; OECD, 2007). Regional and local governments and agencies are also frequently involved in cluster development (Sölvell *et al.*, 2003; Observatory of European SMEs, 2002). The sum of these national and local initiatives represents a significant overall policy effort and it is important to respond to and guide this political interest and the initiatives it is spawning.

Potential threats to cluster performance

Despite the success of many clusters, their continued contribution to economic growth cannot be taken for granted. There are many examples of previously strong clusters that have been unable to reposition themselves and have declined severely in the face of strong international competition or technological and market change, such as the shipbuilding industry in Glasgow, UK, or the New England golf equipment cluster in the USA (OECD, 2001; Porter, 1998).

The market and competitive threats are intense today with the economic crisis. A number of authors refer to a cluster life cycle, with clusters being born, growing, moving to maturity and declining (Bergman, 2008; Maggioni and Riggi, 2008; Huggins, 2008). As with the product life cycle, it is nonetheless possible for clusters to escape this destiny with appropriate renewal strategies and adaptation to change. Rabelloti *et al.* (2009) show how many of the famous Italian industrial districts are adapting to changes in competitive pressures by evolving in their sectoral specialisations, developing new internationalisation and innovation strategies and developing new forms of enterprise organisation. Nonetheless other case study evidence and the experience of clusters that declined or collapsed in the past suggest that clusters can sometimes become locked into their core technologies, reinforced by infrastructures such as social capital and research and training activities that fit those core technologies, leading to decline as outside technologies and markets change over time. Policy makers should therefore be aware of the potential competitive weaknesses of clusters and have some understanding of how lock-in may occur and what facilitates diversification. The speed of reaction to such problems needs to accelerate in current conditions.

Costs of clusters

The economic benefits of clusters are commonly highlighted. However, there are also potential economic and social costs that should be taken into account in considering whether and how to support clusters. One of the most obvious costs is increasing congestion as clusters grow, in the form of busy roads, queuing and rationing of public services and infrastructure such as schools, hospitals and leisure facilities, rising house prices and encroachment of buildings on green spaces. This has led some local government authorities, such as those involved in the planning of the Oxford and Cambridge clusters, to constrain growth or disperse activities to other areas. Another potential cost is that cluster growth may increase social inequalities in host localities as a core group of workers emerges with high pay whilst another segment of society is excluded from core jobs. Whilst all residents suffer from the increased congestion, residents benefit to different degrees from the increased incomes generated by cluster growth. Policy makers are therefore called on to find ways of managing congestion and spreading benefits. Some very interesting local initiatives have emerged to help manage these problems. For example, the OECD identified a series of social innovations in the Silicon Valley and San Francisco Bay area that have emerged in response to rapid growth pressures, including PolicyLink's "Equitable Development Zone" in the City of Richmond and California Futures Network's promotion of planning policies focused on developing existing, underutilised urban centres rather than encouraging urban sprawl (OECD, 2004).

Learning from policy experimentation

In recent years, the number of cluster promotion initiatives has increased, leading to a situation today where many types of policy approach coexist, varying in their precise goals, activities and intensities. Approaches range from leaving the clustering process to the market, building critical mass through inward investment and infrastructure initiatives, supporting science-industry linkages and creating formal networks among cluster enterprises for joint sales and purchasing and other types of co-operation. The co-existence of many approaches presents an opportunity for the policy maker to learn about policy options.

At the same time, the fact that cluster policies have often arisen in a bottom-up way from local initiatives suggests scope for identifying and spreading the most effective approaches. At present, robust evaluations of the economic impact of cluster initiatives are rare, reflecting the complexity and multifaceted nature of cluster policies relative to single policy instruments, and this evaluation gap needs to be addressed. The scope to learn from other places nonetheless exists in a more qualitative way, with the proviso that the initiatives implemented as a result of such exchange of good practices should be properly evaluated and adjusted in line with the results.

One of the most important lessons from cluster experience to date is that policy should not seek to build clusters in places where the necessary critical mass and conditions do not exist, corresponding to the "wishful thinking" clusters of Enright (2003) (policy driven clusters that lack not only a critical mass, but also any particular source of advantage that might promote organic development). This is much less likely to be a problem in the case of major clusters with significant history but caution should be applied when seeking to apply the lessons to very small clusters.

A framework for policy co-ordination

A further reason for focusing on clusters is because they offer potentially useful frameworks for co-ordinating the range of different actors involved in delivering economic policies relevant to cluster development on the ground and better fitting them to real local needs. Taking a cluster perspective allows the policy maker to consider together and at the same time all the aspects of the local environment and firm capacities that influence the competitiveness of key export sectors. This lends itself very well to the development of integrated strategies that bring together a range of approaches and measures – such as inward investment, entrepreneurship, training, innovation and infrastructure – around key local goals, hence increasing the potential for achieving synergies, reaching real priorities and mobilising local stakeholders. Within clusters, the potential is very strong for bringing together policy activities more productively by analysing systemic needs and developing collaborations between stakeholders.

Clusters and the economic crisis

Recent years provided very favourable conditions for cluster development, reflecting strong global economic growth. From 2008, however, a global economic crisis has provided radically different conditions. The crisis is destroying firms and jobs, making finance difficult to obtain for investment and stifling appetites for risk. As concentrations of entrepreneurship and innovation, clusters are likely to be hard hit. At the same time, however, the cost and productivity advantages that clusters have built up over time are likely to provide firms in clusters with an important advantage in surviving the crisis relative to isolated firms. Thus clusters and cluster policies can be seen to be important in both times of growth and recession. Furthermore, if policy makers can make the right decisions about investments to be made now, those that will improve the competitive environment in clusters rather than simply seeking to subsidise everyday firm operations, they are likely to lay the foundations for future economic and employment growth.

How do clusters function?

It is critical to understand how clusters function in order to assess the appropriate role for policy. In one sense, this is not straightforward because there are several competing explanations of the cluster phenomenon (Asheim, Cooke and Martin, 2006) whilst it has already been highlighted that the specific nature and functioning of clusters varies. At the same time, however, there is considerable overlap between explanations and fairly wide agreement about the fundamental foundations of cluster performance.

The core argument goes back to Marshall's assertion that there is a triad of external economies in operation within clusters, namely "thick" labour markets, specialised input suppliers and knowledge spillovers (Marshall, 1920; Belussi and Caldari, 2009). Following Marshall, clusters are widely seen to generate innovation and productivity benefits because they facilitate:

- A division of labour in the labour market, enabling firms to obtain specialised labour with lower search and training costs and higher productivity and workers to obtain higher wages and increased opportunities to specialise and move firms.

 CLUSTERS, INNOVATION AND ENTREPRENEURSHIP – ISBN 978-92-64-04442-5 – © OECD 2009

- A division of labour in intermediate supplier industries, providing customer firms with local access to specialised suppliers of materials and components, finance, marketing and business services and so on, enabling specialised input providers to exploit greater internal economies of scale, reducing search and monitoring costs for both parties, and reducing transport costs.
- Knowledge spillovers, such as transfers of knowledge of market opportunities and technologies among agents in the various firms and research organisations located within clusters through a range of formal and informal links.

Duranton and Puga (2004) give a more differentiated account of the types of externalities that are likely to explain the locational advantages of clusters. However, the key feature of both accounts is that the important externalities are favoured by geographical proximity, thus explaining the phenomen on of clustering. There is certainly evidence of a significant localisation of knowledge spillovers, reflecting the tacit nature of some parts of knowledge (Jaffe *et al.*, 1993; Audretsch and Feldman, 1996; Anselin *et al.*, 1997) and the greater division of labour in the product and labour markets of clusters is also clear to see.

These externalities drive productivity growth, permitting profit and wage growth on the one hand or price reductions and output and employment growth on the other. This may set into motion a positive cumulative causation process. Thus the higher rewards to firms, entrepreneurs and workers that the higher productivity of the cluster offers may attract new factors and activities in terms of new firm starts, growth of existing firms, inward investment, in-migration of labour, growth of research and training centres and so on, thus increasing cluster mass and productivity in a cycle of growth that is likely to come to an end only when increased cluster productivity advantages are matched by greater cluster costs as wages, land prices and other factor costs rise and congestion arises.

Clearly, the case for clusters rests on the idea that they offer favourable environments for high innovation and productivity and there is significant empirical evidence pointing to the existence of such benefits. Aharonson *et al.* (2004), for example, show how clustering affects biotechnology firms' innovativeness, taking firms with similar characteristics and contrasting the performance of those located in clusters with those not located in clusters. They found that clustered firms are eight times more innovative than geographically remote firms, with the largest effects for firms located in clusters that are strong in their own specialisation. Biotechnology may be a sector for which clustering has particular benefits because it relies heavily on the transfer of tacit knowledge. Nonetheless, other studies taking a range of technology specialisations show that cluster firms tend to be more productive in innovation than other firms (Audretsch and Feldman, 1996; Baptista and

Swann, 1998; Breschi, 2008; Lublinski, 2003; Porter, 2003; Wallsten, 2001; Weterings and Boschma, 2006). There is also empirical evidence linking clusters to other economic benefits including increased firm productivity (Madsen *et al.*, 2003; Szforzi, 1990), entrepreneurship (Feldman, 2001; Guiso and Schivardi, 2007; Rocha and Sternberg, 2005; Rosenthal and Strange, 2005), local employment growth (Audretsch and Dohse, 2004; Fingleton *et al.*, 2008; Wennberg and Lindqvist, 2008) and local wages (Brenner and Gildner, 2006; Porter, 2003; Wheaton and Lewis, 2002).

A debate that has not yet been fully resolved, however, concerns whether the important externalities in a cluster occur within particular industries, and are therefore favoured by sector specialisation (*i.e.* localisation economies or Marshall-Arrow-Romer externalities) or whether they occur across industries and are favoured by sector diversity (*i.e.* urbanisation economies or Jacobsien externalities). If they are the latter then it is simply city size that matters. If they are the former then it is the concentration of firms in the same sector that counts, corresponding better to the conception of clusters usually applied in the policy world. There may also be a middle road, in the form of related variety, referring to externalities among activities in different industries but which share competence and knowledge-bases (Boschma, 2009; Asheim, 2008). Empirically, the relative importance of localisation as compared with urbanisation economies has not yet been pinned down (Glaeser, 2000; de Groot *et al.*, 2008), but clearly an important element of localisation economies is required for traditional cluster thinking to hold.

There are other views of cluster functioning and how economic benefits are generated that go beyond the Marshallian externality view outlined above. Particularly influential is the work of Porter (1998) on the competitive advantage of nations, regions and cities. Porter's analysis focuses on four inter-related cluster advantages in his famous "diamond", namely: specialised factor inputs (such as human resources, capital resources, physical infrastructure and university research institutes), related and supporting industries (such as capable suppliers and related industries), sophisticated and demanding customers (that press firms to improve and provide insights on future market needs) and a favourable investment climate and vigorous local rivalry. This represents a broader view of the driving forces of cluster competitiveness than the Marshall story, in particular in its emphasis on the benefits of a competitive spur to productivity. Others have focused on notions of innovative milieu (Ratti *et al.*, 1997; Camagni, 1991), local innovation systems (Asheim and Gertler, 2005; Cooke, 2001, Cooke *et al.*, 2004) and other sources of localised increasing returns. Nonetheless, externalities are usually at the centre of theoretical explanations of cluster performance.

A difficult issue within all this, and one with important potential policy implications, is whether formal networks and collaborations are required to achieve the anticipated benefits of clusters, or whether the benefits can be achieved simply through co-location and normal business transactions. For example, studies of high-technology European clusters found that knowledge transfer largely occurs through labour mobility amongst enterprises, company spin-offs and acquisitions of enterprises by others in the cluster rather than by formal exchanges (Keeble and Wilkinson, 2000). This might suggest that cluster policy should focus mainly on building cluster size rather than interactions. Others place much stronger emphasis on the role of social capital in clusters and suggest that governments may be able to create environments that favour the emergence of social capital and co-operations among firms and other agents in clusters (Storper, 1997; Putnam, 2000).

Furthermore, entrepreneurship is clearly a critical, if under studied, process in cluster development and one that is tightly linked to innovation. A large part of cluster success is usually attributed to relatively strong innovation performance. However, innovation requires entrepreneurs to carry new ideas into exploitation. Thus successful innovation in clusters is likely to be accompanied by new firm formation, spin-outs and small firm growth. To favour this, clusters need to provide appropriate conditions for successful new business development, such as an entrepreneurial culture, access to clients, access to capital, and access to exploitable knowledge in existing firms and research organisations (Mills *et al.*, 2008).

What is the role of policy in clusters?

The fundamental rationale for policy intervention in clusters is to facilitate the exploitation by firms and workers of potential local external economic benefits from input-output linkages and knowledge spillovers and hence increase productivity and generate growth.

One mechanism through which policy may facilitate exploitation of positive externalities in clusters is by stimulating growth in cluster mass, for example by providing information to potentially mobile firms and workers on cluster locations and specialisations, offering incentives for investment in clusters and supporting the provision of housing, business premises and other infrastructure in clusters. This approach creates the conditions for the potential realisation of positive externalities by increasing the co-location of agents. If firms and workers are encouraged to locate in clusters then productivity within the cluster may be expected to increase. There is a national benefit if that increase outweighs reductions in productivity in the places that agents relocate from.

A second mechanism is intervention to encourage collaborations among agents in clusters, for example through incentives for joint research projects, commercialisation of public research, creation of spin-off companies, development of science and innovation parks, creation of local knowledge exchange networks and upgrading of suppliers to assist them in gaining contracts with larger firms. This type of approach assumes that co-location on its own is not enough and that explicit local collaborations are required to realise the potential benefits of clusters. In principle it does not reduce productivity outside the cluster. In fact increased innovation would be expected to spill over to other places.

If the market itself were to generate optimal clustering and cluster interactions then such policy intervention would be counterproductive. However, there are a number of reasons for thinking that the market will produce suboptimal outcomes.

1. *Non-appropriability of externalities.* Clusters are argued to be the sites of positive externalities. However, whilst firms and workers may be attracted to clusters to gain from externalities generated by others, they are not compensated for the positive externalities they generate themselves. For example, if a firm trains skilled labour which then moves on to another cluster firm or imparts tacit knowledge about market opportunities to other firms, it is not directly compensated. It may therefore be expected that firms and workers will engage in too little clustering without some form of public intervention to help clusters emerge and grow.
2. *Lack of location information.* Firms and workers may lack information on existing or potential cluster locations in which their productivity could be enhanced. If this is the case then agents may choose suboptimal locations unless policy makers can find a way to transmit information on the location and specialisation of high productivity local clusters.
3. *Lack of co-ordination of agents.* Although firms benefit from co-location in a cluster, market mechanisms may not be sufficient to co-ordinate the co-location process. Even if agents are aware of the potential benefits of locating near to others, it may not be clear where they should locate to realise these benefits since the eventual outcome depends on a set of decentralised decisions. Thus eventually perhaps no cluster will emerge, or a cluster may emerge in one place and not another. Since the outcome is difficult to predict and agents face costs in relocating there can therefore be a blockage on cluster emergence. A lack of trust among agents may also act as a barrier to the development of potentially beneficial collaborations within clusters such as research and development collaborations and joint purchasing or marketing initiatives. The general difficulty is that no individual agent is likely to be sufficiently recompensed for the costs of co-

ordinating the establishment of a cluster and the interactions within it. If an agent undertakes promotion work to enhance the visibility of a cluster and attract new agents, all present and future players in the cluster are likely to benefit, whether or not they contribute to the promotion work and there is usually no private sector mechanism available to ensure that all beneficiaries contribute to the costs. This suggests that the public sector can have an important role to play in "seeding" the emergence of clusters and as an "honest broker" of collaborations.

4. *Restrictions on mobility*. Barriers to the geographical mobility of firms and workers are also likely to impede the emergence of clusters. It might be expected that firms and workers will be drawn to locate in clusters because of the greater profitability and real wages expected, at least up to a size where the effects of an increasing number of competitors or congestion start to force firms to disperse (Krugman, 1991, Baldwin *et al.*, 2003). However, barriers to the mobility of agents, for example in housing, land and labour markets, may prevent the optimum level of clustering from emerging without public action to address mobility problems.
5. *Generic market failures*. A number of well-known market failures affect entrepreneurship and innovation in general and are also likely to affect clusters. Typical problems include financing for new and growing firms, development of science and technology infrastructure, research and training. As with any market failures, appropriate public intervention is called for as long as the public sector can achieve better outcomes than the market. In the case of cluster policy, the key point is that there are likely to be certain failures that are specific to particular clusters in their scale or nature, and hinder the emergence of potential cluster specialisations. For example, finance, skills, research and infrastructure might be required for a particular activity such as biotechnology or information technologies, suggesting the need for policy solutions that are adapted to the needs of each cluster.
6. *Negative externalities and distributional problems*. Another type of policy problem that may occur in clusters is the possibility that cluster growth may lead to significant congestion costs because agents do not take into account the negative externalities of their activities for others. Policy should therefore price the congestion or take second-best measures to disperse activity or improve infrastructure in the cluster to remove the congestion. Furthermore, the market is not good at addressing potential distributional problems in clusters in the form of the inequalities that may emerge between rich and poor residents. Policy intervention of some kind may be needed to address these distributional questions, whether at national or local level.

There has been strong debate about the role of policy in clusters (Asheim, Cooke and Martin, 2006; Caniels and Romijn, 2005; Cooke, 2002; Hospers *et al.*,

2008; Peck and Lloyd, 2008). However, arguments based on the need for policy intervention to address the types of market failures identified above have strong theoretical and empirical foundations. It could be argued that once clusters are operating successfully the public sector should withdraw since firms should then have internalised the location advantages started through the policy stimulus to co-location and collaboration. In practice, however, continued public sector support may be required for activities that are subject to continued market failures such as the provision of education, training, seed and venture finance and pre-commercial research appropriate to the needs of the cluster.

There is also an emerging school of thought that focuses on evolutionary aspects of cluster development and the adaptability of clusters and their firms to technological, competitive and market demand shocks (Bathelt *et al.*, 2004; Bergman, 2008; Boschma, 2009; Cooke, 2009; Hassink, 2005; Martin and Sunley, 2006; Zucchella, 2006). A key message from this thinking is that clusters must avoid over-specialisation and lock-in by branching from time to time into new directions involving the reorientation of existing activities to new technologies and markets, creating new combinations of activities and exploiting related variety opportunities. It is often argued that policy also has a role to play here in facilitating access to new knowledge via "global knowledge pipelines", or alliances with global technology and market partners, the attraction of foreign direct investment and highly skilled labour and the promotion of innovation and entrepreneurship to support the emergence of more variety within clusters.

However, the potential problems with cluster policies should also be recognised and taken into account in decisions about whether and how to support clusters. Firstly, new theory from spatial economics suggests that policy assessments need to look wider than clusters themselves if they are to establish the full economic impacts of cluster policies, since whilst there may be positive impacts on the competitiveness of the clusters receiving policy support, they may be accompanied by negative impacts on the competitiveness of other areas (Potter, 2009). Indeed, cumulative causation processes are likely to be involved in which the strongest clusters grow at the expense of others. Therefore, whilst local governments and development agencies will wish to capture the maximum cluster activity to their area, at national level it must be recognised that there can be only a small number of major clusters. This suggests an important governance issue requiring co-ordination between national and local policies.

Secondly, cluster policies are essentially a type of "picking winner" approach. In addition to market distortion concerns, they are therefore associated with the usual problem of lack of information on current and future conditions to support governments in the choice of sectors that will grow successfully. There are a number of examples of past cluster policy interventions that have failed because of changes in markets, technologies or competition or because the nature of the intervention was not appropriate. Hospers *et al.* (2008) refer to this problem as one of "backing losers" rather than "picking winners". A related problem is the potential for "institutional capture", whereby public authorities become too close to cluster stakeholders and start to follow their agenda too closely risking high policy deadweight and distortion.

Thirdly, supporting clusters increases the degree to which host economies are specialised in cluster activities. If those clusters then decline, and if it takes time for other industries to grow up locally and for workers to shift to the new industries, then the host locality will suffer from out-migration and reduced incomes and increased unemployment for those who remain.

Finally, clustering clearly should not be seen as the sole economic development solution for innovation and growth. There is a wide range of evidence pointing to the major importance of many other factors such as research and development, human capital formation, entrepreneurship, public infrastructure and so on (OECD, 2003, 2009). The major priority is to strengthen these factors nationally and not solely in a few concentrations, notwithstanding the national importance of certain clusters.

There are therefore some important reasons for policy intervention but also some potential pitfalls. The best policy advice then is to ensure that cluster policies meet real needs on the ground. The following chapters examine cluster development processes in selected major clusters, the policy challenges they face and how policy is responding. The case studies provide important insights for all those involved in developing clusters.

References

Aharonson, B., J. Baum and M. Feldmann (2004), "Industrial Clustering and the Returns to Inventive Activity: Canadian Biotechnology Firms", 1991-2000, Working Paper 04-03, Danish Research Unit for Industrial Dynamics.

Anselin, L., A. Varga and Z. Acs (1997), "Local Geographic Spillovers between University Research and High Technology Innovations", *Journal of Urban Economics*, Vol. 42, pp. 422-448.

Asheim, B. (2008), "Regional Environment for Innovation and Entrepreneurship", in *OECD LEED Entrepreneurship and Local Innovation System Reviews: The Case of Cantabria*, OECD LEED Programme, Paris, pp. 155-187.

Asheim, B. and M. Gertler (2005), "The Geography of Innovation: Regional Innovation Systems", in J. Fagerberg, D. Mowery and R. Nelson (eds.), *The Oxford Handbook of Innovation*, Oxford University Press, Oxford, pp. 291-317.

Asheim, B., P. Cooke and R. Martin (2006), "The Rise of the Cluster Concept in Regional Analysis and Policy: A Critical Assessment", in B. Asheim, P. Cooke and R. Martin (eds.), *Clusters and Regional Development: Critical Reflections and Explorations*, Routledge, London, pp. 1-29.

Audretsch, D. and D. Dohse (2004), "The Impact of Location on Firm Growth", CEPR Discussion Paper 4332, Centre for Economic Policy Research, London.

Audretsch, D. and M. Feldman (1996), "R&D Spillovers and the Geography of Innovation and Production", *American Economic Review*, Vol. 86, No. 3, pp. 630-640.

Baldwin, R., R. Forslid, P. Martin, G. Ottaviano and F. Robert-Nicoud (2003), *Economic Geography and Public Policy*, Princeton University Press, Princeton.

Baptista, R. and P. Swann (1998), "Do Firms in Clusters Innovate More?", *Research Policy*, Vol. 27, pp. 525-40.

Bathelt, H., A. Malmberg and P. Maskell (2004), "Clusters and Knowledge: Local Buzz, Global Pipelines and the Process of Knowledge Creation", *Progress in Human Geography*, Vol. 28, No. 1, pp. 31-56.

Belussi, F. and K. Caldari (2009), "At the Origin of the Industrial District: Alfred Marshall and the Cambridge School", *Cambridge Journal of Economics*, Vol. 33, pp. 335-355.

Bergman, E. (2008), "Cluster Life Cycles: An Emerging Synthesis", in C. Karlsson (ed.), *Handbook of Research on Cluster Theory*, Edward Elgar, Cheltenham, pp. 114-132.

Boschma, R. (2009), "Evolutionary Economic Geography and its Implications for Regional Innovation Policy", OECD LEED Programme mimeo, OECD, Paris.

Brenner, T. and A. Gildner (2006), "The Long-Term Implications of Local Industrial Clusters", *European Planning Studies*, Vol. 14, No. 9, pp. 1315-1328.

Breschi, S. (2008), "Innovation-specific Agglomeration Economies and the Spatial Clustering of Innovative Firms", in C. Karlsson (ed.), *Handbook of Research on Innovation and Clusters*, Edward Elgar, Cheltenham, pp. 167-192.

Camagni, R. (1991), *Innovation Networks: Spatial Perspectives*, Belhaven Press, London.

Caniels, M. and H. Romijn (2005), "What Drives Innovativeness in Industrial Clusters? Transcending the Debate", *Cambridge Journal of Economics*, Vol. 29, pp. 497-515.

Cooke, P. (2001), "Regional Innovation Systems, Clusters and the Knowledge Economy", *Industrial and Corporate Change*, Vol. 10, No. 4, pp. 945-74.

Cooke, P. (2002), *Knowledge Economies: Clusters, Learning and Co-operative Advantage*, Routledge, London.

Cooke, P. (2009), "Economic Development Policy as an Evolutionary Envisioning Process", OECD LEED Programme mimeo, OECD, Paris.

Cooke, P., M. Heidenreich and H.-J. Braczyk (2004), *Regional Innovation Systems: The Role of Governance in a Globalized World*, Second Edition, Routledge, London.

De Groot, H., J. Poot and M. Smit (2008), "Agglomeration Externalities, Innovation and Regional Growth: Theoretical Perspectives and Meta Analysis", University of Waikato, Working Paper in Economics 01/08, Waikato, New Zealand.

Duranton, G. and D. Puga (2004), "Micro-foundations of Urban Agglomeration Economies", in J. V. Henderson and J.-F. Thisse (eds.), *Handbook of Regional and Urban Economics: Cities and Geography*, Elsevier, Amsterdam, Vol. 4, pp. 2063-2117.

CLUSTERS, INNOVATION AND ENTREPRENEURSHIP – ISBN 978-92-64-04442-5 – © OECD 2009

Enright, M. (2003), "Regional Clusters: What We Know and What We Should Know", in J. Bröcker, D. Dohse and R. Soltwedel (eds.), *Innovation Clusters and Interregional Competition*, Springer Verlag, Heidelberg, pp. 99-129.

Feldman, M. (2001), "The Entrepreneurial Event Revisited: An Examination of New Firm Formation in the Regional Context", *Industrial and Corporate Change*, Vol. 10, pp. 861-891.

Fingleton, B., D. Igliori and B. Moore (2008), "Employment Growth in ICT Clusters: New Evidence from Great Britain", in C. Karlsson (ed.), *Handbook of Research on Innovation and Clusters*, Edward Elgar, Cheltenham, pp. 79-106.

Glaeser, E. (2000), "The New Economics of Urban and Regional Growth", in G. Clark, M. Feldman and M. Gertler (eds.), *The Oxford Handbook of Economic Geography*, Oxford University Press, Oxford, pp. 83-98.

Guiso, L. and F. Schivardi (2007), "What Determines Entrepreneurial Clusters?", EUI Working Paper, ECO, 2007/48, European University Institute, Florence.

Hassink, R. (2005), "How to Unlock Regional Economies from Path Dependency? From Learning Region to Learning Cluster", *European Planning Studies*, Vol. 13, No. 4, pp. 521-535.

Hospers, G.-J., F. Suatet and Desrochers (2008), "Silicon Somewhere: Is there a Need for Cluster Policy?", in C. Karlsson (ed.), *Handbook of Research on Innovation and Clusters: Cases and Policies*, Edward Elgar, Cheltenham, pp. 430-446.

Huggins, R. (2008), "The Evolution of Knowledge Clusters: Progress and Policy", *Economic Development Quarterly*, Vol. 22, No. 4, pp. 277-289.

Iammarino, S. and P. McCann (2009), "Local Innovation Systems: Identification and Policy Analysis", paper prepared for the OECD LEED Programme, OECD, Paris.

Jaffe, A., M. Trajtenberg, and A. Henderson (1993), "Geographic Localization of Knowledge Spillovers as Evidenced by Patent Citation", *Quarterly Journal of Economics*, Vol. 108, pp. 557-98.

Keeble, D. and F. Wilkinson (2000), *High-Technology Clusters, Networking and Collective Learning in Europe*, Ashgate, Aldershot.

Krugman, P. (1991), "Increasing Returns and Economic Geography", *Journal of Political Economy*, Vol. 99, No. 3, pp. 483-499.

Lublinski, A. (2003), "Does Geographic Proximity Matter? Evidence from Clustered and Non-Clustered Aeronautic Firms in Germany", *Regional Studies*, Vol. 37, No. 6 and 7, pp. 453-467.

Madsen, E., V. Smith and M. Dilling-Hansen (2003), "Industrial Clusters, Firm Location and Productivity: Some Empirical Evidence for Danish Firms, Aarhus School of Business", Working Paper 03-26.

Maggioni, M. and M. Riggi (2008), "High-Tech Firms and the Dynamics of Innovative Industrial Clusters", in C. Karlsson (ed.), *Handbook of Research on Innovation and Clusters: Cases and Policies*, Edward Elgar, Cheltenham, pp. 54-78.

Markusen, A. (1996), "Sticky Places in Slippery Space: A Typology of Industrial Districts", *Economic Geography*, Vol. 72, No. 3, pp. 293-313.

Marshall, A. (1920), *Principles of Economics*, 8th edition, Macmillan, London.

Martin, R. and P. Sunley (2006), "Path Dependence and Regional Economic Evolution", *Journal of Economic Geography*, Vol. 3, No. 1, pp. 395-437.

Mills, K., E. Reynolds and A. Reamer (2008), "Clusters and Competitiveness: A New Federal Role for Stimulating Regional Economies", paper from the Metropolitan Policy Program at Brookings, Brookings Institution, Washington DC.

Observatory of European SMEs (2002), *Regional Clusters in Europe*, Luxembourg: Office for Official Publications of the European Union.

OECD (2001), *Devolution and Globalisation: Implications for Local Decision-Makers*, OECD, Paris.

OECD (2003), *OECD Growth Study*, OECD, Paris.

OECD (2009), *Going for Growth 2009*, OECD, Paris.

OECD (2004), "Social Innovation and the New Economy: Evidence from the Silicon Valley and the San Francisco Bay Area", OECD Local Economic and Employment Development Programme mimeo prepared by Karen Chappel, OECD, Paris.

OECD (2005), *Business Clusters: Promoting Enterprise in Central and Eastern Europe*, OECD, Paris.

OECD (2007), *Competitive Regional Clusters: National Policy Approaches*, OECD, Paris.

Paniccia, I. (2006), "Cutting through the Chaos: Towards a New Typology of Industrial Districts and Clusters", in B. Ahseim, P. Cooke and R. Martin (eds.), *Clusters and Regional Development Critical Reflections and Explorations*, Routledge, London, pp. 90-114.

Peck, F. and C. Lloyd (2008), "Cluster Policies and Cluster Strategies", in C. Karlsson (ed.), *Handbook of Research on Innovation and Clusters: Cases and Policies*, Edward Elgar, Cheltenham, pp. 393-410.

Porter, M. (1998), *On Competition*, Harvard Business School Publishing, Boston, Massachusetts.

Porter, M. (2003), "The Economic Performance of Regions", *Regional Studies*, Vol. 37, No. 6 and 7, pp. 549-578.

Potter, J. (2009), "Evaluating Regional Competitiveness Policies: Insights from the New Economic Geography", *Regional Studies*, Vol. 43.

Putnam, R. (2000), *Bowling Alone: The Collapse and Revival of American Community*, Simon and Shuchter, New York.

Rabellotti, R., A. Carabelli and G. Hirsch (2009), "Italian Industrial Districts on the Move: Where are they Going?", *European Planning Studies*, Vol. 17, No. 1, pp. 19-41.

Ratti,R., A. Bramanti and R. Gordon (1997), *The Dynamics of Innovative Regions: The GREMI Approach*, Ashgate, Aldershot.

Rocha, H. and R. Sternberg (2005), "Entrepreneurship: The Role of Clusters. Theoretical Perspectives and Empirical Evidence from Germany", *Small Business Economics*, Vol. 24, pp. 267-292.

Rosenthal, S. and W. Strange (2005), "The Geography of Entrepreneurship in the New York Metropolitan Area", *Federal Reserve Bank of New York Economic Policy Review*, Dec. 2005, pp. 29-53.

Sölvell, Ö., G. Lindqvist and C. Ketels (2003), *The Cluster Initiative Green Book*, Ivory Tower, Stockholm.

Storper, M. (1997), *The Regional World. Territorial Development in a Global Economy*, Guildford Press, New York.

Szforzi, F. (1990), "The Quantitative Importance of Marshallian Industrial Districts in the Italian Economy", in F. Pyke, G. Beccatini and W. Sengenberger (eds.), *Industrial Districts and Inter-firm Co-operation in Italy*, International Institute of Labour Studies, Geneva, pp. 75-107.

Wallsten, S. (2001), "An Empirical Test of Geographic Knowledge Spillovers using Geographic Information Systems and Firm-Level Data", *Regional Science and Urban Economics*, Vol. 31, No. 5, pp. 571-599.

Wennberg, K. and G. Lindqvist (2008), "How do Entrepreneurs in Clusters Contribute to Economic Growth?", Stockholm School of Economics Working Paper Series in Business Administration, 3, Stockholm School of Economics, Stockholm.

Weterings, A. and R. Boschma (2006), "The Impact of Geography on the Innovative Productivity of Software Firms in the Netherlands", in P. Cooke and A. Piccaluga (eds.), *Regional Development in the Knowledge Economy*, Routledge, London, pp. 63-83.

Wheaton, W. and M. Lewis (2002), "Urban Wages and Labour Market Agglomeration", *Journal of Urban Economics*, Vol. 51, pp. 542-562.

Zucchella, A. (2006), "Local Cluster Dynamics: Trajectories of Mature Industrial Districts between Decline and Multiple Embeddedness", *Journal of Institutional Economics*, Vol. 2, No. 1, pp. 21-44.

ISBN 978-92-64-04442-5
Clusters, Innovation and Entrepreneurship

Chapter 2

The Micro-nanotechnology Cluster of Grenoble, France

by
Jonathan Potter*
Centre for Entrepreneurship, SMEs and Local Development, OECD

This chapter analyses the emergence and establishment of the micro-nanotechnology cluster in Grenoble, which has been recently labelled the Minalogic "pôle de compétitivité" by the French government. Grenoble is a particularly good example of a cluster that emerged "endogenously" from a critical mass of human and social capital but with strong support from targeted initiatives by the public sector. The chapter demonstrates the role that public authorities can play in supporting a cluster and the importance of establishing strong private-public and public-public partnerships. It shows how innovation has occurred in a collaborative but highly competitive environment. It highlights the importance of human and social capital and the need for "co-ompetition" as key ingredients of an internationally successful cluster.

* The author would like to thank Gabriela Miranda (OECD), Jérôme Domens (DCASPL) and Erwan Pouliquen (DCASPL) for their support in preparing this chapter together with the authors who contributed papers that provided material for the chapter, namely Bernard Pecqueur (Institut de Géographie and Université Joseph-Fourier de Grenoble), Alain Dupré and Eric Fricotteau (Institut national de la statistique et des études économiques, Direction régional, Rhône-Alpes), Jérôme Domens and Erwan Pouliquen (DCASPL), Jean-Marie Rousseau (Consultant, Brussels), Pier Paolo Saviotti (Université Pierre Mendès-France, Grenoble) and David Catherine (Grenoble École de Management).

Introduction

In 2005, the French Government adopted an industrial strategy based on the introduction of "competitiveness clusters" (*pôles de compétitivité*). For the purpose of this policy, clusters have been defined as "the combination, within a given geographic area, of businesses, training centres and public or private research facilities working in partnership to generate synergies in connection with innovative joint projects having the requisite critical mass for international visibility".[1]

To this end, the Inter-ministerial Committee on Territorial Planning and Development (CIADT) conferred the label "cluster" on 67 geographically circumscribed areas and earmarked EUR 1 500 million over the years 2006-08 to underwrite collaborative research and development projects (EUR 1 400 million), finance other kinds of projects such as innovation platforms (EUR 100 million) and support cluster management and leadership groups (EUR 36 million). A second tranche of EUR 1 500 has been allocated for a new phase of the "pôles de compétitivité" programme for the period 2009-11. Businesses belonging to a cluster also qualify for tax exemptions on their research and development (R&D) activities, up to the ceilings authorised by the European Union, and for reduced social security contributions on the wages of their R&D staff.

The Minalogic cluster of Grenoble-Isère has been designated as one of seven "global clusters" in France because of their strong international renown. The title Minalogic is short for *MIcro NAnotechnologies et LOgiciel Grenoble-Isère Compétitivité*, and, as its name in French suggests, it is a cluster specialising in research, development and product design in the sectors of micro- and nanotechnologies and embedded software. Collaborative research projects among firms and research organisations in the cluster are financed under the "pôles de compétitivité" programme by the Business Competitiveness Fund (FCE) of the central government's Directorate-General for Enterprise. This budget is drawn from a pool of finance provided by the government's Single Inter-ministerial Fund, a number of public agencies (the National Research Agency, the Industrial Innovation Agency and OSEO, the French innovation and SME agency) and the Caisse des Dépôts et Consignations, the government's investment bank, and is supported by tax exemptions and reductions in social security contributions for R&D activities. In parallel, a

 CLUSTERS, INNOVATION AND ENTREPRENEURSHIP – ISBN 978-92-64-04442-5 – © OECD 2009

cluster management group brings together actors in the cluster in order to stimulate collaborative research projects, disseminate the results of research and promote the cluster internationally.

Nature and evolution of the cluster

In 2007, the core of the cluster comprised approximately 40 000 jobs in microelectronics and computer science, including 35 000 in businesses and 5 000 in public-sector research, within a metropolitan area with a population of approximately 600 000. However, the Grenoble area has long been a centre for microelectronics related activity. To understand how the cluster developed, one must look not only at events surrounding the recent award of its "pôle de compétitivité" label, but also at the history of the "Grenoble ecosystem" – an ecosystem that to a large extent sprang up informally and spontaneously and then gradually gathered strength and became more formalised.

This section begins by introducing the Grenoble site's leading players, then focuses more closely on the activities of firms belonging to the formal structures of the Minalogic "pôle de compétitivité" and then sets out a summary of major recent developments at the Grenoble site.

Leading players

The leading players in the cluster, aside from government, come from the realms of industry, training and research.

Industry

How the Grenoble cluster's activities are catalogued depends largely on the chosen perimeter in terms of sectors of activity and geography. Focusing on metropolitan Grenoble and the computer science, software, microelectronics and nanotechnologies industries yields a total of 280 businesses in 2007, most (94 per cent) of which had fewer than 200 employees. The main characteristics of these firms are as follows:

- They were *created recently* (most being fewer than ten years old), suggesting thorough renewal of the cluster over the past decade.
- A large proportion results from *spin-offs* or incubation involving a research laboratory.
- Activities are *heavily complementary* and revolve around a small number of core technologies with extensive potential applications.

Among the largest enterprises were five firms in the computer technology industry, each employing over 500 people, the biggest being Hewlett Packard, Bull and Sun Microsystems. In microelectronics and nanotechnologies, three firms employed more than 500 people, the largest of which was STMicroelectronics, with a payroll of 6 000 in Grenoble. In the other main specialism of the cluster, embedded systems-on-a-chip software, no firm had more than 500 employees.

Education

Higher education in Grenoble enjoys a good reputation internationally. One out of every six students is a foreigner, and the proportion rises to 25 per cent in doctoral programmes. There are four universities in Grenoble: the Institut National Polytechnique Grenoble (a technological university for engineers and researchers with 5 200 students); the Université Joseph-Fourier (a science and technology school with 17 500 students); the Université Pierre-Mendès France (a humanities and social studies university with 19 000 students); and the Université Stendhal (for the study of humanities, languages and communication with 7 500 students). A relatively pioneering policy initiative, "Grenoble Universités", that is now underway at the site should ultimately lead to the creation of a single university consolidating the activities of today's four institutions. In addition, the Grenoble École de Management, which is run by the Chamber of Commerce and Industry, has a yearly enrolment of 2 500 students.

Research

There are four national research institutes in areas related to the core activities of the cluster:

- *CEA-Grenoble*: The French atomic energy agency (CEA) has an important operation in Grenoble that is strongly involved in transfers of knowledge and know-how to industry. With a staff of 3 600, including 2 700 CEA employees, it encompasses 115 laboratories, including personnel from Leti.
- *Leti*: Leti is a laboratory of the CEA and is one of Europe's largest centres of applied research in electronics employing around 1 000 people. Leti works very closely with firms and research and education actors in the cluster, and is a major player in the cluster's flagship projects such as Alliance Crolles 2 and Minatec.
- *CNRS Alps delegation*: The CNRS Alps delegation is an operation of the national centre for scientific research comprising over 1 600 officials in 9 research institutes. Since 2001 its laboratories in Grenoble have registered 90 patents, signed 46 licences and produced some 25 start-up companies.
- *INRIA Rhône-Alpes*: Established in December 1992, the Rhône-Alps unit of the French National Institute for Research in Computer Science and Control (INRIA) is one of the Institute's six research facilities in France. It employs some 500 people focused on software and systems-on-a-chip technologies.

CLUSTERS, INNOVATION AND ENTREPRENEURSHIP – ISBN 978-92-64-04442-5 – © OECD 2009

In addition, there are four major international research institutions in the cluster: the Institut Laüe-Langevin, the European Synchrotron Radiation Facility, the European Molecular Biology Laboratory and the Grenoble High Magnetic Field Laboratory.

There are also some important private research activities and public-private research partnerships in the cluster, as illustrated by the following:

- *Alliance Crolles 2*: This project represents an important partnership agreement among three major companies in the cluster – Freescale, NXP Semiconductors and STMicroelectronics. Its aim is to pool research projects and set up a customised research centre in the town of Crolles. It has involved a joint capital investment of roughly EUR 2 billion and research and development outlays of EUR 1.5 billion. It is the largest industrial investment in France in over ten years and has received substantial government support.
- *NanoSmart Center*: In 2006, a company called Soitec joined forces with the Leti research laboratory to launch the NanoSmart Center – a world class centre of excellence in advanced materials. The centre, which is expected to employ some 200 researchers at the Bernin site (near Crolles) by 2010, is focusing on research into new applications for Soitec's "Smartcut" technology.
- *bioMérieux*: The Mérieux Corporation is a major multinational company in *in vitro* diagnostics present in some 130 countries. It inaugurated its worldwide research centre in molecular biology and microsystems in Grenoble in April 2006.
- *Schneider Electric*: The Schneider Electric company has based one of its major research laboratories in Grenoble, employing 1 400 R&D staff and accounting for 20 per cent of its global research budget. The company chairs the governance structure of the Minalogic "pôle de compétitivité".

A portrait of the enterprises in the Minalogic pôle de compétitivité

A concentration of technological activities

In 2006, the enterprises formally included in the Minalogic "pôle de compétitivité" structure, *i.e.* those receiving support from the national "pôles de compétitivité" programme, comprised 46 establishments and approximately 10 000 employees. Some of the member establishments belonged to groups that operate other facilities in Grenoble, representing at the same date 82 establishments in the Grenoble *arrondissement* that belonged to cluster enterprises, with 13 800 employees.

Over its 46-establishment perimeter, the Minalogic cluster is highly concentrated geographically. Apart from one establishment based in the Rhône-Alpes region and another in Haute-Savoie, the remaining 44 establishments were all located in the *département* of Isère, 42 of which were in the *arrondissement* of Grenoble.

The establishments can be grouped into five sectors. Three of the sectors, accounting for 61 per cent of the establishments and 56 per cent of the employees, may be considered to form the core of the cluster's activity: component manufacturing (4 000 employees), software development (920 employees) and engineering and technical inspection (380 employees). The cluster also has a downstream sector ("areas of application"), whose very broad spectrum of activities covers 16 establishments and 4 500 employees, and a support sector ("administrative logistics") provided by two of the cluster's establishments.

Table 2.1 breaks down the number of establishments and jobs for each of the cluster's five sectors of activity in 2006, at various geographical levels. Taking just the activities in which the cluster establishments were specialised, just over one-in-five employees in Isère work for a cluster establishment, as do one-in-fifteen Rhône-Alpes employees and just under one-in-one hundred in France as a whole. In the component manufacturing sector in particular, Minalogic accounts for a consequential share of the total, since its establishments employ one-half of Isère's workers in the sector, one-third of the region's and one-in-twelve nationwide.

Table 2.1. **Number of establishments and workers belonging to the Minalogic cluster's five sectors of activity, 2006**

	Minalogic		Isère		Region		France	
	Ests	Workers	Ests	Workers	Ests	Workers	Ests	Workers
Component manufacturing	12	4 437	61	8 667	165	11 826	1 200	55 125
Software development	9	919	891	6 133	4 118	19 681	42 887	185 630
Engineering and technical inspection	7	377	1 152	5 410	5 462	24 522	47 550	212 042
Administrative logistics	2	47	1 171	2 269	6 407	18 813	56 423	189 437
Fields of application	16	4 467	875	22 439	4 613	78 680	39 043	546 058
Total	**46**	**10 247**	**4 150**	**44 918**	**20 765**	**153 522**	**187 103**	**1 188 292**

Source: INSEE Rhône-Alpes.

Over the decade 1996-2006, the number of people working in the Grenoble cluster in establishments engaged in these activities has increased by 13 per cent. For the Minalogic grouping's core activities alone, job growth was a good deal greater, rising by 163 per cent in component manufacturing, by 68 per cent in software development and by 44 per cent in engineering and technical inspection.

As indicated in Table 2.2, the Minalogic establishments have a substantial proportion (44 per cent) of managers and higher intellectual occupations. Engineering and technical inspection in particular, along with software development, are sectors that employ many managers, and proportionally

CLUSTERS, INNOVATION AND ENTREPRENEURSHIP – ISBN 978-92-64-04442-5 – © OECD 2009

Table 2.2. **Breakdown of Minalogic employees in Isère by socio-professional category and by sector**

Percentage, 2004

SC	Managers and higher intellectual occupations	Intermediate occupations	Office workers	Manual labourers	Total
Component manufacturing	39.0	20.2	3.6	37.2	100.0
Software development	88.0	8.5	3.2	0.3	100.0
Engineering and technical inspection	91.4	7.4	1.2	0.0	100.0
Administrative logistics	24.4	46.3	24.4	4.9	100.0
Fields of application	42.8	36.9	10.5	9.9	100.0
Minalogic	**44.1**	**25.2**	**6.0**	**24.6**	**100.0**

Source: INSEE Rhône-Alpes (Clap, using 8 170 posts).

more so within the cluster (respectively 91 per cent and 88 per cent) than in the *département* as a whole (respectively 43 per cent and 70 per cent).

A special survey was conducted by INSEE, the French national statistical and economic studies agency, for the OECD in 2007 among the 46 establishments in the Minalogic grouping to supplement the statistics available in this area. The following paragraphs set out the main findings of this survey.

Sharp growth

Two-thirds of the cluster's aggregate turnover is derived from its core activities in the five sectors cited above, the growth of which has been extremely sharp since 2002 (rising by 64 per cent), especially within SMEs (up by 90 per cent). Activity related directly to these activities became predominant between 2002 and 2005, rising over the period from 50 per cent to 63 per cent of the total, possibly because of greater profitability, increased capital investment, the cluster's greater visibility or heightened co-operation.

At the same time, other activities are carried out by Minalogic establishments, generating turnover which on the whole has declined since 2002 (by 5 per cent). In SMEs, these activities are still a source of brisk growth (up by 66 per cent).

Highly skilled employees

Three-quarters of the cluster's jobs are in establishments that belong to a group. Because of the strong R&D positioning of the cluster establishments within their groups, four out of ten jobs are filled by researchers and design engineers, while only 1.5 out of 10 relate directly to production. While the employment structure has remained stable since 2002 in multi-establishment firms, independent establishments seem to have substituted production jobs for some of their research and design engineering positions. This would seem to suggest that some of their R&D projects have been completed and that production is now on-stream. For the cluster as a whole, dependent employment has risen by an aggregate 6.4 per cent since 2002, with the sharpest gains benefiting production engineers (+35 per cent), administrative staff (+10 per cent) and researchers and design engineers (+8 per cent).

Widespread and outward-reaching subcontracting

Three-in-four establishments say they are party to subcontracting contacts, involving an average of 40 partners. A large majority of these dealings are with firms outside the cluster, dispelling the idea of a self-sufficient cluster focused exclusively on internal ties, especially since a majority of the subcontracting involves the cluster's core specialisms rather than support services. The most common pattern is that of a buyer within the cluster and a supplier outside it.

Nevertheless, the pattern seems to differ between SMEs and establishments belonging to groups. Group establishments are usually buyers, assuming the role of supplier in only 10 per cent of cases. Indeed, some groups devote all of their activity to R&D and engage in no production at all. For their part, however, 45 per cent of SMEs are suppliers, most commonly to purchasers outside the cluster.

A concentration of collaborative R&D

More than three out of four establishments report at least one co-operative link within the cluster, with either another establishment in the Minalogic grouping, a research laboratory or a university. In all, establishments have forged an average of six co-operative links, which in over 90 per cent of cases involve innovation and R&D, and which rarely extend to other functions in the corporate value chain. In comparison, for industry as a whole (Haag, Raulin and Souquet, 2004), ties are formed most frequently "around the manufacturing activity, in procurement and production". This finding highlights the cluster's originality, as regards both R&D and co-operation.

For the most part, the most strategic co-operative ties forged with other businesses in the cluster tend to be fairly recent and concentrated in the realm of product design (accounting for 80 per cent of co-operation situations). One-third of the time, the co-operation leads to product production. The main purpose of co-operation seems to be to remedy a lack of in-house labour and capital and in most cases does not appear to have required any further allocation of resources or of staffing in particular. Co-operative ties within the cluster involve relatively few contracts, only one-in-three establishments having signed a bilateral contract with its partner.

CLUSTERS, INNOVATION AND ENTREPRENEURSHIP – ISBN 978-92-64-04442-5 – © OECD 2009

Ties with laboratories and universities seem to have been in place for longer than ties with other businesses. A large majority involve product design or, to a lesser extent, production. Co-operation with laboratories or universities most commonly involves the supply of human resources. From a contractual standpoint, the preference for relatively informal relationships remains strong, even if subcontracting contracts are used in most cases. Businesses may consider co-operation with laboratories as a way of outsourcing a portion of their R&D. Achieving greater flexibility would seem a more common motivation for co-operation with a laboratory than for dealings with another business. But in most cases establishments look to laboratories or universities for the same reasons they co-operate with other cluster establishments – because they lack the requisite in-house skills or equipment.

A high level of innovation

Establishments in the cluster are extremely innovative: over 80 per cent of them are currently developing new products or processes and the same proportion of establishments had brought new products to market since 2002. Nonetheless, SMEs in the cluster showed something of an innovation deficit as compared with groups. Slightly more groups marketed new products or processes; moreover, a larger share of SMEs had registered no patents or licences since 2002 (30 per cent, *versus* only 18 per cent for group subsidiaries). On the other hand, if projects currently being developed are taken into account, the gap seems narrower, with SMEs and groups reporting comparable levels of innovation. A learning curve for SMEs having embarked upon such activities more recently may explain the shrinking differential.

Impediments to innovation

The greatest impediments to further innovation would appear to be a lack of financing, uncertainty over demand, the cost of innovation and the difficulty finding partners for such projects. Furthermore, while the level of innovation seems thoroughly comparable between SMEs and groups, the perceived difficulties involved may differ. Contrary to all expectations, lack of funding and the costs of innovation are more often seen as impediments by groups than by SMEs, while one might assume that it would be less difficult for an establishment belonging to a group to line up funding than for an SME. This might be explained by a higher average cost of innovation projects in larger groups. On the other hand, the difficulty in finding collaboration partners for innovation is perceived much more as an impediment to innovation by SMEs than by establishments backed by groups.

Grenoble's changing circumstances

Analysis of INSEE statistics confirms a decade of job growth in the sectors of activity of the cluster. As shown in Figure 2.1, in the *arrondissement* of Grenoble, employment has increased over the past ten years by 163 per cent in component manufacturing, 68 per cent in software development and 44 per cent in engineering and technical inspection. Over the same period, aggregate job growth in all sectors combined was 13 per cent. These figures show a positive employment trend in the recent evolution of the Grenoble cluster, led by the core cluster sectors.

Figure 2.1. **Employment trends in the Grenoble arrondissement**

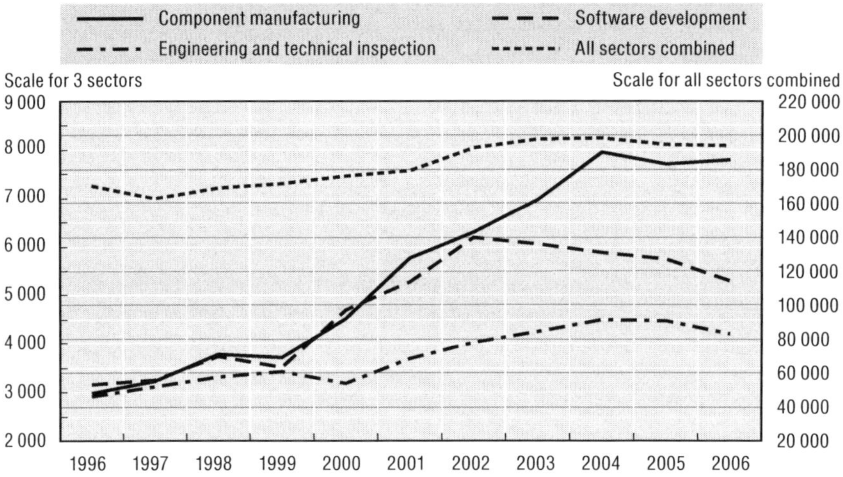

Source: INSEE Rhône-Alpes (Sirene).

Over a longer period, an important feature of the Grenoble cluster is the continuous evolution of its activities in line with broader technological change, which has enabled the cluster to adapt and grow. Table 2.3, taken from Pecqueur (2007), shows how the Grenoble cluster has updated its markets, its form of organisation and its policy strategy over time. This has involved combining and recombining a number of basic technological staples (materials, particle physics, computer science, microchips etc.) in various configurations, from the era of hydro-electric power, electrochemistry and electrometallurgy in the 1950s to the present era of nanotechnology. As in other global clusters, this capacity to evolve along with technical progress may be considered one of the keys to Grenoble's success.

The site's current activity revolves around micro- and nanotechnologies and is supported by four technological pillars: nanotechnologies, biotechnologies, software and new energy technologies. These pillars benefit from a stream of

Table 2.3. **Long-term trends in the Grenoble cluster**

Period	Market	Organisation	Strategy
1950-60s	• Development of hydroelectric power, electrochemistry and electrometallurgy.	• Large "Fordist" enterprises.	• Adapt traditional industry to available new technologies.
1970s	• Investment in research and innovation gradually transforms production. Traditional products lose share.	• Arrival of multinational firms and government investment in research. Emergence of specialised SMEs.	• Intensification of the structural adaptation process.
1980s	• Traditional products on the decline. • Branching out towards emerging products: computers and electronics.	• Externalisation and intense specialisation of SMEs. • Higher-skilled labour. Higher average wages.	• New capabilities and reinforced initiatives by local authorities (with the decentralisation of 1982-83).
1990s	• Reduced production of material goods. • Grenoble recognised as a technology cluster. • Large firms and networks of SMEs turn to exports. • Denuclearisation and emergence of bio- and nanotechnologies.	• Installation of a "core and ring" system, with efficient, research-oriented SMEs gravitating around multinationals.	• Declining government investment (national policy) forces the site to seek new solutions for maintaining its leadership (Minatec).
2000s	• Nanotechnological orientation as a generic theme, with flourishing of technological hybrids (Jean Therme's "three-leaf clover").	• Multi-party governance with three types of competition: academia (INPG/UJF), research (CEA/CNRS/INRIA), industry (ST/ Schneider, etc.).	• Site policy revolving around co-operation mechanisms: alliances, clusters, etc.

Source: Pecqueur (2007).

innovations resulting from a solid base of basic research undertaken locally or translated locally from research undertaken elsewhere. As shown in Figure 2.2, nanotechnologies appear as innovative and specific "meta-technologies" combining the site's accumulated knowledge and know-how.

Figure 2.2. **Technologies and industries in the Grenoble cluster today**

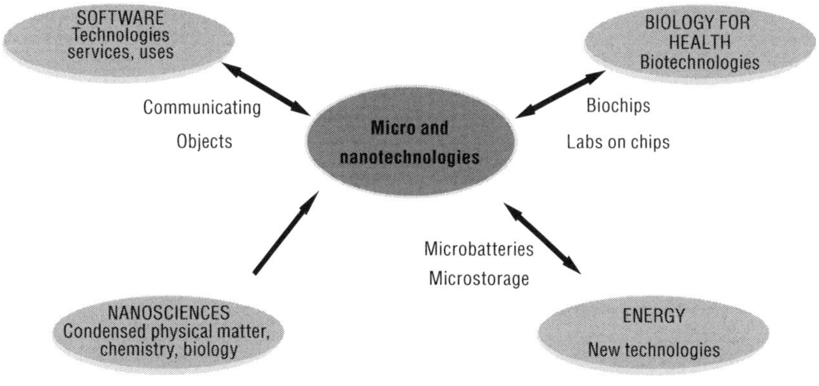

Source: AEPI 2006.

Success factors

It is possible to identify five key factors behind the success of the Grenoble cluster in the form of some influential structural projects, the close involvement of local public authorities in the development of the cluster, and the availability of strong human capital, social capital and public research. Each key factor is discussed in turn below.

Structural projects

A number of major structural projects have been undertaken in recent years with large scale investments by the public sector and local partners in business, research and education. These appear to have contributed significantly to the cluster's development. Three of the most important of these recent flagship projects are Minatec, Alliance Crolles 2 and Minalogic.

Minatec

Minatec is a major European centre of expertise in micro- and nanotechnologies. It comprises:

- An education and training platform in a purpose-developed building housing 1 000 engineering students, 120 instructors, a number of doctoral schools and 500 trainees in an ongoing-training centre.
- A research platform comprising some 40 laboratories and 1 200 researchers equipped with clean rooms.
- A commercialisation platform, in a building that houses growth phase start-ups and joint research teams from public-sector laboratories and large private sector corporate R&D operations.

The project was initiated by one of the major research institutions in the cluster, the CEA-Grenoble, and one of the major education and research institutions, the technical university of the Institut National Polytechnique Grenoble. It also had strong backing from local and territorial authorities. The total investment amounts to roughly EUR 1 billion over a five-year time frame.

Alliance Crolles 2

This project is a collaborative industrial development programme in nano-electronics that started up in 2002 between STMicroelectronics, Philips-NXP and Freescale. The three companies have pooled their financial and human resources in order to co-operate on R&D issues while continuing to compete in the "downstream" part of the value chain. The project is an outgrowth of longer-standing collaboration between STMicroelectronics and Philips-NXP involving teams of engineers, which by 2006 had come to involve thousands of people working together on targets for 2010.

CLUSTERS, INNOVATION AND ENTREPRENEURSHIP – ISBN 978-92-64-04442-5 – © OECD 2009

The co-operation has not taken the form of a joint venture insofar as each partner remains independent in managing human resources and marketing products. The three Alliance Crolles 2 partners have opted jointly for simple processes: a core unit for processing and integrating new staff, to which each firm can add its own particular requirements. The project also enjoys substantial collaboration with other cluster stakeholders, including research institutions, university laboratories in the region and, above all, CEA-LETI. The national and local governments have also contributed to investment in the project's infrastructure and research.

Minalogic

Minalogic is the name of the cluster support initiative supported by the French government in Grenoble through its "pôles de compétitivité" programme. Minalogic supports research and development activities in the creation, development and production of products and solutions in the realm of intelligent miniaturised services for industry. The Minalogic cluster's principal role is essentially that of a facilitator of the creation and financing of common research and development projects. It is co-managed by a grouping of the key enterprises, research and education institutions and public authorities in the cluster. These organisations had already worked co-operatively before the cluster label was awarded, but the Minalogic initiative aims to promote this collaboration further.

Minalogic had 78 members as of January 2007, including 48 businesses (33 SMEs), 10 research centres and universities, 14 territorial authorities, six economic development organisations and one associated private investor. At the core of Minalogic's governance structure is its tripartite general meeting and board of directors (see Table 2.4). All members of the cluster are represented in these bodies. In addition, a leadership group made up of four standing members implements decisions taken by the board of directors in respect of Minalogic's daily operations.

Table 2.4. **Minalogic governance structure**

	Tripartite general meeting	
• Large groups	• Research	• Territorial authorities
• SMEs	• Education	
	Board of directors	
• Chairperson	• Members (2)	• Secretary
• Treasurer		
• Member (1)		

The Minalogic initiative can obtain "labels" for projects that permits them to be submitted for calls for projects potentially leading to funding. As of January 2007, Minalogic had obtained labels for 79 projects. Following the first two calls for projects, in February 2006 and May 2006, seven Minalogic projects were selected, attracting aggregate subsidies of EUR 36.4 million, financed by the State (EUR 23.1 million), the Rhône-Alpes region (EUR 4.5 million), the *département* of Isère (EUR 4.5 million) and Isère municipalities and municipal communities (EUR 4.3 million).

Involvement of local authorities

Local authorities, and specifically the Rhône-Alpes region and the Isère General Council, have provided consistent and long-term support for the cluster's development in order to support local economic development. This support can be seen in a number of ways: land-use planning, involvement in the Minalogic board of directors, financial aid for R&D projects and investment in the cluster's major structural projects (Crolles 1, Minatec, Alliance Crolles 2 and Minalogic).

Human capital

Another factor conducive to the cluster's growth is the availability, for businesses and research organisations, of skilled labour in its fields of research and innovation. The cluster's human capital is enhanced by the presence of a number of institutions of higher education, including four universities with aggregate enrolment of 60 000 students in 2006, 10 per cent of whom were foreigners. These institutions are a major source of skilled labour for cluster businesses and research organisations. Universities and research centres, including the CEA and Leti, are also major sources of knowledge which contribute to the cluster's innovations in a variety of fields. Table 2.5 shows the number of employees, researchers and graduates in the cluster's two leading fields of innovation: computer technology/software; and micro-/nanotechnologies and electronics.

Table 2.5. **Human capital in the cluster's leading fields of innovation, 2006**

	Businesses (jobs)	Public research (jobs)	Higher education
Computers and software	12 000	1 500	2 200 graduates per year
Micro-/nanotechnologies and electronics	21 700	3 000	1 200 graduates per year
Total	33 700	4 500	6 800 students 3 000 graduates (without double counting)

Source: AEPI, 2006.

Another factor behind the positive development of the Grenoble cluster lies in its ability to attract internationally renowned scientists (including the arrival of Professor Louis Néel, winner of the Nobel Prize for Physics in 1970, followed by that of Louis Weil and Paul Louis Merlin), as well as highly skilled workers and enthusiastic entrepreneurs. These people, and their ideas and projects, have been successfully incorporated into the cluster's dynamic.

Social capital

The cluster benefits from a collaborative culture in which actors from different institutional sectors work together relatively easily, indicating the existence of strong social capital. There are thus frequent contacts between the research sector and industry and between public and private stakeholders. These collaborations are aided and encouraged by geographical and institutional proximity. A large share of applied research within the cluster is conducted in partnership between public research and firms, thereby encouraging exchanges of important information between researchers and entrepreneurs as projects, which in some cases may be interdisciplinary, are implemented.

Stakeholders from the public sector are closely involved in both the research projects pursued within the cluster and the creation of infrastructure to meet the cluster's needs. One of the factors in the cluster's success is the large number of collaborative projects involving actors from different backgrounds (public institutions, industry, research sector) as well as the mobility of individuals between these different areas. Many public decision-makers, for example, have previously held posts in industry, academia or the research sector, which undoubtedly allows them to find a better match between public aid and actors' needs.

Public research

One of the strongest drivers of competitiveness of the cluster is the high concentration of public research resources in universities and research centres. This represents a critical mass of public research which can often provide a basis for private sector projects seeking to develop practical applications. Grenoble has a population of 15 000 researchers working in universities and research centres (11 000 in public research and 4 000 in private research), including the CEA, LETI, CNRS and INRIA.

Role of SMEs

The extent to which small and medium-sized enterprises (SMEs) help to drive the development of the Grenoble cluster, particularly in terms of new knowledge and technology, is not easy to evaluate. A survey of establishments in the Minalogic cluster grouping undertaken for the OECD showed that the cluster's SMEs were as innovative as the large firm establishments. However, their role in subcontracting arrangements differed from that of group subsidiaries in that they devoted more time (45 per cent) to performing work for contractors, in most cases from outside the cluster, than subsidiaries did (10 per cent). Furthermore, SMEs participated in both the governance of Minalogic (General Assembly and Board of Governors) and labelled research projects. Table 2.6 indicates the number of Minalogic projects funded by the Business Competitiveness Fund of the Directorate-General for Enterprises in its first three calls. The share of SMEs receiving such funding is rising and has increased from 1.4 per cent for the first two calls for proposals to 21.9 per cent at the third call for proposals.

Table 2.6. **Share of SMEs in Minalogic projects which received public funding**

Calls for proposals	Number of projects selected	Total number of partners	Total aid funding (EUR million)	% of SMEs (< 250 employees)	% of major groups	% of public laboratories
1st (Feb. 2006) and 2nd (May 2006)	7	29	36.4	1.4	53.2	45.4
3nd (Dec. 2006)	7	28	21.5	21.9	36.6	41.4

Source: Directorate-General for Enterprise/*provisional figures.

Impact of the cluster on entrepreneurship and employment

The Grenoble cluster appears to have had a positive impact on enterprise creation. According to the survey of establishments in the Minalogic grouping in 2006, 40 per cent had started up since 2002. Furthermore, among the cluster's SMEs, more than one in three had started up in 2006, which would suggest that there is a strong dynamic for enterprise creation within the cluster.

To date the Minalogic activities unit has provided assistance in the creation of three companies and start-ups. In addition, a private investor (a venture capitalist) joined Minalogic as an associate member at the end of 2006, which should help to finance start-ups in the future. However, while Minalogic has acquired cross-holdings in the project committees of the GRAIN business incubator and the GRAVIT research consortium[2] in order to provide support for and increase the pace of enterprise creation, no formal mechanism has been set up in Minalogic to directly encourage the creation of enterprises. In particular, there is no specific mechanism to finance or provide seed capital for start-ups. The creation within Minalogic of an activity dedicated to developing enterprise creation would go a long way towards ensuring the closer integration of SMEs.

The Minalogic initiative also seems to have had a positive impact on the growth of SMEs already in place. The SMEs within the cluster openly acknowledge that the cluster aids their development in a variety of ways.

 CLUSTERS, INNOVATION AND ENTREPRENEURSHIP – ISBN 978-92-64-04442-5 – © OECD 2009

Firstly, the labelling of a project by Minalogic enhances their credibility and increases their chances of receiving funding from other organisations (ANVAR, FCE, venture capital companies, etc.). Moreover, the systematic inclusion of SMEs in co-operative projects, as well as their participation in the governance of Minalogic, offers them the possibility of joining new networks in the Grenoble ecosystem.

Assessing the impact on job creation of public aid to the cluster would require a detailed study. Some useful information can nonetheless be gleaned from an analysis of the impact of the Alliance Crolles projects 1 and 2 on employment in the Isère *département* (a study conducted by the Université Pierre-Mendès-France and the Reverdy Associés consultancy). This study estimated the number of jobs created indirectly (*i.e.* in the rest of the value chain for various types of production) and induced (*i.e.* induced by the income redistributed within the territory by such direct or indirect activities) by the Alliance Crolles projects 1 and 2 as well as the jobs created directly on the site (in the companies STMicroelectronics, NXP and Freescale).

In 2004, the number of jobs created directly by the Crolles site was estimated to amount to 2 500 for Crolles 1 and almost 1 100 for Crolles 2. Some 660 jobs were estimated to have been indirectly created by Crolles 1 and 2 among service suppliers, 90 among materials suppliers and 740 among equipment suppliers. A further 470 jobs were indirectly created in research centres, primarily the Leti. Complementing these job creations, there were a further 1 300 jobs at the STMicroelectronics unit in Grenoble relating to projects upstream and downstream of Crolles 1 and 2. The number of induced jobs over this period would seem to be around 4 000 in all areas of activity related to household consumption (education, health, trade, administration) and construction. Overall, the estimate reveals that the presence of three partners of the alliance in Crolles accounts for approximately 10 900 jobs in the Isère *département*. This survey estimates ratio of jobs created indirectly or induced to those created directly at the Crolles site to be around 3:1.

SMEs sometimes find it difficult to enter into co-operative relations with groups or research institutes in the cluster, which may well act as a dampener on growth in activity and employment within the cluster. The difficulty of finding partners was one of the specific problems raised in SME replies to the survey questionnaire. In addition, there exists a large asymmetry between SMEs and group subsidiaries in their satisfaction with public policies aimed at bringing actors closer together through co-operation, with very high satisfaction in the case of large firm establishments but much lower satisfaction reported by SMEs. Two main problems in this area were identified in the course of the interviews. Firstly, some of the most negative SMEs see Minalogic as a means of redistributing public research funding to major groups, as a result of which SMEs are apparently less inclined to participate fully in such research projects. Moreover, even though the collaborative projects labelled by Minalogic each include at least one SME, some SMEs felt that the projects proposed were designed more with the interests of major groups in mind.

Barriers to cluster development

The Grenoble cluster is the key driver of growth of the Grenoble economy. Nevertheless, there are certain barriers that could constrain the future development of the cluster, which should be taken into account in future policy decisions.

One obstacle signalled by local stakeholders concerns a reduction in the number of researchers and technicians. One-fifth of the cluster's employment is dedicated to research. However, because of relatively low salaries in the public sector, young graduates from engineering schools seem to be increasingly targeting employment in large enterprises rather than research laboratories and universities. In addition, a gap in the number of technicians with respect to engineers has been reported, suggesting that there may be a lack of appropriate technical or vocational training. For example, there is no training of technicians for clean rooms. Although firms themselves could provide the appropriate technical training more might be done by universities and other public actors.

The scale and quality of public infrastructure has also been signalled by local stakeholders as a constraint to the development of the cluster. The Grenoble agglomeration has seen a rapid increase in population in recent years, including the inflow of a significant number of professionals from other parts of France and overseas. This has increased the demand for schools, transport, housing and other infrastructure. Furthermore, a complication in the case of Grenoble is that the city is surrounded by mountains, which represent a natural frontier limiting its size and growth. In these circumstances, in-migration has a strong impact on housing and property prices and road congestion. Although the local authorities are investing in public transport, and notably the tram, the necessary investment is very large and will not resolve the problems on its own.

Another problem felt locally is that of "delocalisation" or relocation of establishments to competitor regions, particularly in the east of Europe where there is qualified labour at lower cost. Investment in innovation is the major response to this challenge. At the same time, however, it is argued locally that public administration in some countries is able to finance research in Grenoble's core research activities more rapidly and at a greater scale than is the case in France, thus threatening the technological lead of the Grenoble cluster.

Some local actors have also raised concerns about the limits on the flexibility of public research funding under the "pôles de compétitivité"

CLUSTERS, INNOVATION AND ENTREPRENEURSHIP – ISBN 978-92-64-04442-5 – © OECD 2009

programme. The "pôles de compétitivité" legislation aims to concentrate funding on activities in specific geographic locations in order to maximise the efficiency of spending. However, recipients do not have access to financing to develop activities in fields outside of the nominated specialisation of the cluster they belong to, as defined by the programme. Thus the Grenoble cluster is encouraged to dedicate its research to the fields of micro- and nanotechnologies and embedded software without the possibility of major public support to branch into other fields. The risk therefore is that Grenoble could become a "single sector" cluster without the necessary capacity to react to new market needs.

There are also some barriers to entrepreneurship in the cluster. One aspect of this is an insufficient supply of risk capital to support the creation and development of innovative SMEs, which is a problem in France in general and that also appears to affect the Grenoble cluster. Another aspect appears to be a lack of a generalised entrepreneurial culture in the local scientific community and the existence of "invisible barriers" to movement between research and industry.

Finally, some groups are opposed to the development of the cluster for two main reasons. Firstly there is some opposition from local enterprises due to the limited representation of SMEs in the core governance and research and development activities of the Minalogic "pôle de compétitivité" structure which is dominated by large firms and research institutions. Secondly, there is a minority expressing strong ethical objections to some aspects of nanotechnology research undertaken by the core cluster actors. Both groups raise objections that potentially could lead to local opposition to investments in the cluster if they are not addressed appropriately.

Role of policy

As highlighted above, the public authorities have played an important role in promotion of the Grenoble cluster, even though the contribution to the activities of large firms appears to have been greater than the direct contribution to the creation and growth of SMEs. Of course the largest contribution by the public authorities to development of the cluster consists in providing ongoing and capital funding for the four universities and the research laboratories and in the development of public infrastructure in the cluster rather than through initiatives solely designed with the objective of supporting the cluster in mind. However, many initiatives have been designed more specifically to help the cluster evolve. Table 2.7 sets out the main initiatives that enterprises (large firm subsidiaries and SMEs) in the Grenoble cluster recognise as being important for the development of the cluster.

Table 2.7. **Public initiatives to promote the Grenoble cluster**

Programme	Public organisations	Type of contribution
Minalogic cluster	Central government, territorial authorities	Promotion, funding and management of research projects in the micro- and nanotechnology sectors and embedded software (EmSoC – Embedded Systems on Chip).
Alliance Crolles	Central government, territorial authorities, public research institutions	R&D collaboration project in nano-electronics between three firms from the cluster at a site in Crolles.
Minatec	Territorial authorities, public research institutions	Teaching platform, research and applications development of micro and nanotechnologies.
Metis	Territorial authorities	R&D projects aimed at disseminating micro-nanotechnologies to SMEs in traditional sectors such as textiles and paper.
Agence d'Étude et de Promotion de l'Isère (AEPI)	Territorial authorities	Prospecting for new investors, partners and markets abroad; assistance to investors.
"Forum 4i"	Territorial authorities, Chamber of Commerce and Industry	Raising venture capital for innovative enterprises in a forum comprising venture capitalists and investors in the Grenoble project.
Grenoble Angels	Chamber of Commerce and Industry	Association of entrepreneurs who share their experience and network and offer financing for innovative start-ups.
Grenoble Alpes Incubation (GRAIN)	Central government, Chamber of Commerce and Industry, research institutions	Business incubator which finds and provide facilities for plans to create enterprises to exploit public research
Support for hiving off activities from research laboratories	Central government	Incentives to exploit innovations arising from basic research conducted by public research organisations such as the CEA and INRIA
OSEO	Central government	Assistance to SMEs through support for innovation, funding of investments and provision of financial guarantees.
1999 Act on innovation and research	Central government	Facilitating the mobility of research staff to industry; promoting partnerships between public research and enterprises; improving the tax regime for innovating enterprises; improving the legal framework for innovating enterprises.
Tax Credit for Research	Central government	Tax break for enterprises equal to half the R&D expenditure during the fiscal year, less the average expenditure of the same type during the previous two years.
Status of Young Innovating Enterprise	Central government	Tax breaks for qualifying SMEs less than 8 years old and with R&D expenditure amounting to at least 15 per cent of their charges.
National aid competition for the creation of enterprises and innovative technologies	Central government, European Union	Development of plans to create enterprises to exploit innovative technologies through grants and appropriate support measures.

Policy adaptations over time

The emergence of the Grenoble cluster has prompted the Grenoble public authorities to shift their R&D and economic development policies towards an approach based on partnerships with enterprises and training and research institutions that is aimed at fostering innovation.

CLUSTERS, INNOVATION AND ENTREPRENEURSHIP – ISBN 978-92-64-04442-5 – © OECD 2009

Accordingly, the territorial authorities have worked collectively to coordinate their actions at the levels of the region, *département* and communes to provide infrastructure and financing for research and innovation in collaboration with private sector partners from the cluster. The level of support provided by the territorial authorities in this respect is currently very high and has increased substantially over the past few years, primarily in response to the cluster's growing need to invest in research. The Grenoble Chamber of Commerce and Industry (CCIG) is another actor whose support for innovative enterprises has increased and is now more active than Chambers in many other cities in France. Even though the CCIG's actions in this respect remain limited compared with those of public stakeholders, examples of new innovation initiatives include Espace Entreprendre, to provide advice, training and mentoring to entrepreneurs, and the Grenoble Angels association and "Forum 4i" initiatives to increase access to seed capital.

At national level, the creation of the competitiveness clusters programme also demonstrates a political will to adapt research and innovation policies to the needs of major clusters.

Future policy challenges

Responding to increased international competition

Globalisation has been accompanied by increased competition in the Grenoble cluster's sectors of activities. There is very strong competition from other OECD countries. The United States and Japanese governments in particular are promoting the development of nanotechnologies and information technology by providing funding for both R&D and the subsequent development of applications for research findings. Firms and research institutes in Grenoble are often obliged to make major investments to expand their activities. They seek partnerships with the public sector to support these investments. Public investment aid to meet the needs of the Grenoble cluster must therefore be both timely and flexible as well as on a sufficient scale to keep pace with the competition. In broader terms, it is important to continue to maintain an environment favourable to innovation within the cluster.

Involving SMEs and start-ups in cluster initiatives

Start-ups and SMEs often play a vital role in disseminating innovation within clusters. These small entities enter into the value chain by commercialising the innovations developed by major enterprises and laboratories, thereby allowing the latter to concentrate on their core activity. However, the activities of the Grenoble cluster have been strongly weighted to large enterprises to date.

Promoting the creation and development of SMEs would make it possible to disseminate innovation more widely and increase the critical mass of the cluster, the key to raising its profile to global level. There are, however, a number of factors that act as a brake on achieving this objective. Firstly, researchers are not familiar enough with the process of spinning off activities and are not motivated strongly enough to go and work for, or set up, an SME. In addition, there are no *ad hoc* structures in place in Grenoble to promote the creation of high-tech enterprises on the basis of work performed by universities and research laboratories. The GRAIN business incubator, which is an effective tool for spinning off start-ups from the public research sector, does not collaborate directly with the cluster. Nor does the Grenoble Chamber of Commerce and Industry have a unit specialised in creating new high-tech enterprises.

Moreover, more active collaboration between research bodies and existing SMEs would benefit both sets of actors. It would allow public research bodies to ensure they secure the best possible economic benefits, while at the same time allowing SMEs to obtain the support they need to refine the technologies they develop. However, again there are major barriers to the very principle of such collaboration. Firstly, research institutes have neither the right culture nor sufficient incentives to collaborate with SMEs. Secondly, SMEs are not familiar enough with collaborative projects and exhibit a certain degree of reticence. There would therefore seem to be a need to widely advertise the mutual benefits of co-operation and, in addition, encourage the creation of systems and procedures that will protect and reassure SMEs about issues such as intellectual property. This was one of the issues revealed by the survey of cluster enterprises, which showed that many SMEs had not yet filed a patent or registered a licence.

Improving access of innovating SMEs to sources of financing

Access to financing is often one of the barriers that SMEs face when they are first set up and when they later start to grow. Innovative SMEs in a cluster need both substantial and long-term financing given that the revenues expected in such high-tech sectors are earned much later than the initial investments. Funding from private sources would seem to be essential in this area. However, there is still not enough venture capital available in the Grenoble cluster. This deprives innovative SMEs of additional sources of capital for their creation and growth, leaving the burden to fall on public funds. In France, the number of venture capitalists remains very limited, and this is likely to cramp the development of start-ups and innovation within its clusters. The public authorities have a role to play in ensuring growth in this type of private financing, particularly in the high-tech sector. With regard to the co-financing of projects by the Minalogic initiative, it is important that the public share of the funding is made available as soon as possible. Public financial aid to SMEs should be capable of rapid mobilisation and should have administrative procedures that are as streamlined as possible.

Increasing the use of SMEs as suppliers to the central hub of the cluster

Job creation also depends, more broadly, on local SMEs that are not directly involved in the cluster's activities. These enterprises can not only play a role as subcontractors or suppliers for enterprises and research laboratories in the cluster, they can also supply consumer services for cluster employees. To maximise the ripple effect on the local economy as a whole, these enterprises must be able to offer products and services that are tailored to the needs of the cluster. It is therefore important that SMEs have reliable information about the type of demand that they may receive from the cluster. SMEs may also need support to introduce any internal process innovations that may be required to meet this demand. In this respect, the survey of cluster establishments stressed the amount of ground that SMEs at the cluster had lost with regard to group subsidiaries in terms of process innovation. The Grenoble Chamber of Arts and Crafts and the Chamber of Commerce and Industry have a role to play in supporting SMEs in this regard and their work would be facilitated by better communication between the cluster and commercial networks.

Increasing co-ordination and visibility of entrepreneurship support

Several very efficient public bodies already offer support for entrepreneurship in the Grenoble cluster. However, entrepreneurs are sometimes not sufficiently aware of the actions of such bodies and it can sometimes be difficult for enterprises to participate in them. Co-ordination is not always efficient and entrepreneurs in the cluster are sometimes faced with a variety of different contact persons, dealing with the same issues, in the various institutions tasked with aiding SMEs. It might therefore be helpful to increase the visibility of the public systems put in place to aid SME creation and development and make it easier to gain access to them.

Maintaining and increasing the number of researchers and technicians

The success of the Grenoble cluster is largely based on its large pool of human capital in the fields of research and technology. However, over the past few years, the number of researchers in Grenoble has been falling. The introduction of more vigorous initiatives to attract more researchers to laboratories and enterprises in the cluster would be helpful. Likewise, there are not enough technicians to meet the cluster's requirements, a shortage that, in particular, is slowing down the creation of clean rooms.

The universities lack room for manoeuvre to tailor their courses and teaching programmes to the current and future skill needs of the cluster. The public authorities could help to remedy this shortcoming by giving universities greater room for manoeuvre, as well as earmarked funding for such training courses.

Promoting labour mobility

The exchange of knowledge between industry and research is critical to innovation in a cluster. In this respect, collaborative research projects involving co-operation between actors from industry and researchers are a very effective means of exchanging knowledge. This type of exchange appears to be highly productive within the Grenoble cluster.

In contrast, the mobility of workers between industry and the research sector is still not sufficiently developed in Grenoble. There are substantial constraints on the mobility of workers, whether it be from large firms to SMEs or between industry and the research sector. The public sector has a part to play in encouraging such mobility, notably by providing support and assistance for employees wishing to leave a major enterprise to either join or create an SME. Universities and engineering schools could also help to facilitate exchanges between researchers and entrepreneurs, as well as mobility between industry and the research sector.

Striking the right balance between diversity and specialisation in innovation

Clusters often struggle to strike the right balance between exploring new avenues to capitalise on research, which can ensure the branching of the local economy into new areas, and the exploitation of existing avenues of research, which will ensure the excellence of current economic development. If too much energy is devoted to exploring new avenues for development the exploitation of current research may be hindered in the short-term, and the cluster may well start to decline. On the other hand, if efforts are focused on seeking short-term practical applications for research, the cluster's development may eventually be halted in the longer term.

Grenoble would appear, if anything, to be running the risk of over-specialisation. The State "pôle de compétitivité" programme only finances research and development projects in sectors that are foreseen by the formal regulations governing the activities of the Minalogic initiative, which limits the scope of its projects. This is likely to encourage firms in the Minalogic cluster to focus their activities to micro and nanotechnologies and embedded chip software without adding to their expertise in other fields. The risk of becoming a "single speciality" and "single sector" cluster could jeopardise its international competitiveness and reduce its reactivity to the new requirements of international markets. Actors from the cluster are therefore calling for its sectors of specialisation to be broadened.

On the other hand, greater diversity within Minalogic could well undermine the strength of the fields that are of most importance to the cluster by spreading skills too thinly. In this respect, it would be helpful to open the debate on relaxing the system used to allocate financial aid within a cluster in order to promote innovation in sectors whose expertise complements or mirrors that in core cluster specialisms.

Improving national and international access in terms of transport

Grenoble has a relatively poor transport infrastructure in terms of national and international connections: no express regional shuttle train services between Grenoble and the other economic metropolitan centres such as Lyon, Chambéry and Geneva; lack of a direct high-speed train link from Paris; congested motorways. Making good these weaknesses might enhance the attractiveness of the cluster in the eyes of foreign direct investors and encourage the mobility of engineers. On the other hand, the lack of an international airport does not appear to detract from the city's attractiveness, since Grenoble is located close to two international airports, namely Saint-Exupéry near to Lyon and Geneva.

Continuing and broadening support by national administrations for cluster initiatives

The Grenoble cluster requires investment in research infrastructure, in research and development projects and in infrastructure in the broad sense of the term. Given that the cluster is of national importance and that the economic benefits from growth in the cluster extend well beyond the boundaries of territorial administrations, it is logical to involve national authorities in this investment effort. The Minalogic cluster and the planning contracts between central government and the regions are major expressions of the State's involvement. This effort needs to be maintained over the long-term.

Managing tensions in the real estate and transport sectors

The attractiveness of Grenoble is under threat from its relatively high degree of congestion, the outcome of strong demographic and economic growth in an area that is naturally constrained by the mountains and rivers surrounding the city. This congestion is currently having an adverse effect on transport, making it harder to travel to cluster establishments. Roads are saturated during rush hours and, despite major investment in public transport, difficulties still remain. The tramways in Grenoble only run as far as the municipal boundaries and do not provide access to the major industrial complexes in peripheral areas. The city's physical expansion appears to have reached its limits, which has led to rising property prices fuelled by the arrival of high-earning managers. The creation of new enterprises and the growth of those already in place are therefore subject to the constraints posed by the scarcity of land for industrial development and the difficult task of attracting employees, who have to contend with the high price of accommodation. Overall, the prices of land and property in Grenoble and surrounding areas are starting to hold back growth in the cluster and the saturation of the transport networks risks damaging its attractiveness. New solutions are required to these problems.

Avoiding the risk of social divide

There is a risk that growth in the cluster only favours the income and employment of researchers, engineers and technicians working in the cluster and does nothing for those who live in the city but have no contact with the cluster and none of the qualifications required to find work in the cluster. If the growth and high earnings of the high-tech enterprises in the cluster were to have no positive impact on employment and salaries in the rest of the city, this could be viewed as a failure in terms of economic and social development. Furthermore, the resulting "social divide" might well encourage both the inhabitants of the city and local authorities to resist the future development of the cluster.

To avoid this, wealth-redistributing mechanisms need to be designed. Improving the technological level of SMEs, promoting subcontracting and co-operation with enterprises in the cluster might be potential solutions. It is also important to encourage the training of the local population, particularly in activities with insufficient manpower, in order to improve their employability in the enterprises within the cluster. In addition, it would be helpful to raise the general awareness of the population of the science and technologies developed in the cluster in order to encourage people to pursue careers in the relevant fields. The local tax revenues generated by the cluster's activities should allow the territorial authorities to put in place better facilities to promote the social integration of those members of the population who are furthest removed from the high-earning activities of the cluster.

This risk of "social divide" in Grenoble is linked to the cluster's concentration on R&D. The region has few component manufacturers, for example, in the automobile, avionics, telecoms or biotechnology sectors – in which the cluster's technologies are nonetheless present – which would be capable of providing mass employment. Attracting and anchoring such enterprises in Grenoble would create jobs commensurate with the qualifications of the local population.

Meeting ethical objections to research

The contents and objectives of projects supported by Minalogic are not always very well known to the general public, which can fuel suspicion and concern. The research and development carried out in the cluster relate to advanced fields of technology and may raise certain ethical questions which need to be discussed and addressed. For example, concern may arise over the health and safety of employees working on nanotechnologies, which may be perceived as dangerous products to handle. Further, some applications of the products produced by nanotechnologies – for military use, espionage, etc. – might also be considered to have negative social impacts.

For such reasons, there is a small group of people who are opposed to the activities at Minalogic and who project a negative image of the cluster to the population of Grenoble, who, as taxpayers, are called upon to fund investment in the cluster. The opacity of cluster structures, the difficulty in gaining access to the central site and the lack of information on the objectives and contents of projects undoubtedly help to fuel this trend. An effort needs to be made to communicate and teach people about the cluster's activities and innovations in order to make it easier for the general population to understand the issues at stake. It is also important to respond to the genuine concerns that are raised by taking ethical considerations into account in the design of cluster activities. A number of associations such as *ACROR* (the laboratory run by the *INPG*), and *ACONIT* (*Association pour un conservatoire de l'informatique et de la télématique*) are pursuing interesting initiatives in this respect. Organising virtual visits to laboratories, for example, helps to promote greater scientific awareness and the trickle-down of technology to small traditional enterprises. Such activities have tended to create bridges between the scientific and entrepreneurial culture and the concerns of the general public.

Lessons for other clusters

The experience of Grenoble offers a number of key messages for other clusters.

First, Grenoble demonstrates the potential clusters have to drive economic development. In direct terms, the growth in turnover of core cluster establishments was 64 per cent in the period from 2002 to 2006. The employment growth of the cluster establishments over a ten year period was 13 per cent, rising to 163 per cent in component manufacturing, 68 per cent in software development and 44 per cent in engineering and technical inspection, namely in the three key core sectors of the cluster. In indirect terms there is significant subcontracting, with cluster establishments working with an average of 40 suppliers, mainly outside of the cluster, and induced effects from the local spending of wages and salaries. An estimate of the induced and indirect effects of the Alliance Crolles project suggests a ratio of around three local jobs created for each direct job in a cluster establishment. It is these job and income effects that policy seeks to stimulate. Furthermore, the fact that cluster firms put out significant non-local as well as local subcontracting and that cluster establishments represent a significant proportion of national employment in their specialisms suggests that the cluster plays a national as well as local role in the competitiveness and growth of its industries.

Second, the Grenoble case illustrates the importance of adopting a long-term policy strategy for supporting a cluster and allowing this strategy to adapt, because clusters take time to emerge and the environments they operate in change. The origins of the Grenoble cluster can be traced back at least 50 years to the development of hydro electric power in the Alps and to the establishment of industry-relevant research establishments. Since then, the cluster has exploited various different structural technologies, in line with global technological changes. The capacity of policy, research, education and industry to adapt to new technical and market conditions at each step, whilst maintaining a consistently positive attitude to the development of the cluster, seems to have been critical to Grenoble's success.

Third, the Grenoble cluster has been supported by strong national public investments in education and research. It benefits from a very strong education sector, including four universities and a management school, delivering a large inflow of skilled labour to the cluster. It also benefits from eight national and international research institutes that are relevant to the cluster and form the basis for collaborative research projects and spin-offs in the cluster domains. The importance of this national research and education activity to the development of the cluster implies that national decisions on the location of research and education investments in general should not be taken without regard to locality, but should, as one of the criteria used, take into account the potential of different locations to support clustering.

Fourth, a number of public investments have been made in infrastructures and research and development collaborations in the Grenoble cluster. In particular, national, regional and local government authorities have all worked together to provide support to three flagship projects in collaboration with research and education institutions and the private sector. The Alliance Crolles 2 initiative provided a research facility for pooling the pre-competitive research of three major cluster companies. The Minatec centre provided education and training facilities for 1 000 students and 500 continuous education trainees, research facilities for 40 laboratories and 1 200 researchers and business incubation facilities for spin-off enterprises. The Minalogic "pôle de compétitivité" provides funding for R&D activities and a cluster management organisation for brokerage of collaborative research

 CLUSTERS, INNOVATION AND ENTREPRENEURSHIP – ISBN 978-92-64-04442-5 – © OECD 2009

projects. The cluster's strong current international position should therefore be seen not as simply private sector led, but in part the product of targeted public investments in cluster-specific infrastructure, undertaken in public-public and public-private partnerships. These types of "structuring investment" are also likely to be important to the development of other clusters.

Fifth, the greatest impediments to innovation reported by Grenoble cluster establishments, large and small, are lack of financing, uncertainty over demand, the cost of innovation and difficulty finding partners. Apart from dealing with uncertainty over demand, these are all critical and well-established areas of policy intervention in enterprise and innovation policy in general. The location of enterprises in clusters, however, can give a boost to policies seeking to address these problems by enabling them to concentrate resources and call on synergies with the actions of a range of local partners.

Sixth, the success of the Grenoble cluster in large part reflects the outputs of its pool of very highly skilled labour. An estimated 44 per cent of employees of core cluster establishments are managers or in higher intellectual occupations. Moreover, a key feature of the establishments belonging to large firms in the Minalogic structure is that they serve mainly an R&D and design role within their parent companies. They are therefore specialised within their group in highly skilled work. It is the ability of the cluster to host this specialised R&D and high skill work that would appear to count above all. Public policy can be argued to have played an important role in facilitating access of cluster enterprises and institutions to high-skilled labour through delivering appropriate education and training as well as by providing an attractive residential environment.

Finally, it is clear that there are a number of policy challenges still to meet in Grenoble and it is very likely that a number of them are shared with other clusters. Indeed the challenges still to meet in Grenoble might be thought of as among those that are potentially more difficult to meet or that have been less visible or popular with policy makers. Thus there appears to be an "invisible barrier" between industry and public research and a lack of SME linkages into core cluster activities that potentially constrains knowledge transfers. There is also the issue of whether to support cluster specialisation or diversification in the way that research and development funding and other policy support is provided to the cluster. Policy makers need to take a view on whether the future growth of the Grenoble cluster is likely to be found in hybrid activities that combine existing cluster strengths with new activities, or whether branching into new activities will disperse the effort too far and see a drop in the core critical mass required for cluster success. It has recently become evident that congestion associated with the growth of the cluster also needs to be managed with appropriate transport, housing and land policies and investments, that an emerging social divide needs to be countered by finding ways to spread the benefits and that ethical objections also need to be addressed through greater transparency and dialogue with civil society. Public authorities and their partners in Grenoble are looking to the experience of other clusters in providing inspiration on the most effective approaches for meeting these new challenges. The lesson for other clusters comes from this very effort. An effective cluster policy is likely to be one that is open to information on emerging challenges and how they may be addressed.

Notes

1. As defined in the call for projects by the Office for Territorial Development and Regional Action (DATAR).
2. Grenoble Alpes Valorisation et Innovation Technologique (GRAVIT) is a consortium made up of the CNRS, CEA, INP Grenoble, INRIA, UJF and UPMF as part of a mutualisation programme sponsored by the Agence Nationale de la Recherche for a period of three years (2006-2008). The INP Grenoble is the project co-ordinator and manager.

References

Haag, O., E. Raulin and C. Souquet (2004), "Inter-business Relations" in *Le 4 pages des statistiques industrielles*, SESSI, No. 195, November.

Pecqueur, B. (2007), "Historique du Cluster Grenoblois", paper for the OECD LEED Programme, OECD LEED Programme, Paris.

ISBN 978-92-64-04442-5
Clusters, Innovation and Entrepreneurship

Chapter 3

The High-tech Cluster of Oxfordshire, United Kingdom

by

Helen Lawton-Smith
Managing Director and Director of Research Oxfordshire Economic Observatory and Reader in Management, Birkbeck, University of London

This chapter provides an overview of the transformation of the high-tech cluster in Oxfordshire following industrial restructuring in the UK. The case study illustrates the importance of the entrepreneurial dynamic as a source of talent, spin-offs and critical mass. It highlights the challenges faced by the Oxfordshire cluster in networking among the actors of the "virtuous circle" for clusters, and the ways in which these actors have been involved in order to fulfil the unmet needs of the cluster. The success of the cluster has been underpinned by the attractiveness and self-organisation of the territory. This chapter also outlines the key role played by the core actors of the cluster in tackling environmental challenges and social inequalities.

Introduction

Oxfordshire is a key growth area in the UK economy and is one of Europe's leading centres of enterprise and innovation. In 2002 it received its second Award of Excellence as one of Europe's most innovative regions. The transformation from an old economy based on brewing, agriculture, blankets, motor vehicles and education, to a new mixed economy, in which the cluster of high-tech firms makes an important contribution to the county's prosperity, has been remarkable. It is an example of a sub-region where the economy has undergone a major shift from the commodity producing industries of the twentieth century towards one based more on the potential of the knowledge and information economy. By the start of the new millennium Oxfordshire, and its surrounding region, the South-East, are among the top high-tech locations in Europe. Highlighted in this account of the growth of the Oxfordshire high-tech cluster are on the one hand, the positive impact of national science and innovation policies and entrepreneurial activity and an increasing concentration of talent (Florida, 2002) at the local level; and on the other, the negative impacts of rapid growth. The paper explores the consequences of both in planning policy and in future training provision. It draws extensively on research conducted by the Oxfordshire Economic Observatory, an independent research centre, based in Oxford University.

Nature and evolution of the cluster

From the mid 1980s to the mid 1990s, the Oxfordshire economy underwent a transformation with: the decline of the car industry which had employed 27 000 in the late 1960s but only around 5 000 by 1997; a rapid increase in the formation rate of high-tech firms and the arrival of the R&D departments of a small number of predominantly US and Japanese multinational firms; and a clearly identifiable system of local governance that was not based in the public policy domain. During the 1980s, the high-tech cluster grew from 190 firms employing 2 000 people at the end of the decade (Lawton-Smith, 1990) to some 543 firms employing 19 465 in 1997 (Garnsey and Lawton-Smith, 1998). By 2002, the number of high-tech firms had risen nearly threefold to 1 400 high-tech and employment slightly less than doubled at 36 700, representing about 12 per cent of all employees working in the county (Oxfordshire Economic Observatory, 2003). The most recent estimate is of approximately 3 500 businesses employing around 45 000 in 2004 (Oxford

Economic Observatory, 2007). The definition used in all three studies is based on that developed by Butchart 1987 for the DTI (referred to as "Butchart"), with the latest study using an extended version to reflect the growth of new high-tech sectors. The Butchart definition uses the criteria of R&D expenditure (as a percentage of turnover) and qualified scientists and engineers (as a percentage of all full-time employees) to indicate whether or not a firm be classified as high-tech.

While the early high-tech economy was dominated by manufacturing, the Oxfordshire high-tech cluster has grown into being services dominated. There are far more high-tech service companies than high-tech manufacturing companies and service employment is higher than that in manufacturing overall. The sector with most businesses is computer services, with almost half of all the high-tech companies in the county (635 firms, 45 per cent of companies) which has twice as many companies as technical consultancy and technical testing (22.5 per cent) which is also an important high-tech services sector. The largest manufacturing sector is instruments, medical and optical equipment, followed by biotech/pharmaceuticals. Certain sectors, although important in employment terms, consist of only a small number of companies. For example the motorsport and automotive engineering/design sector accounts for less than 2 per cent of the county's high-tech firms but 7 per cent of its high-tech jobs. The emerging biotech sector has 73 firms but only comprises 5.2 per cent of the county's high-tech firms (Table 3.1). Table 3.2 shows the distribution of employment within the high-tech economy.

Success factors

The main drivers of this rapid growth are enterprising individuals drawn from an increasingly highly skilled workforce, "entrepreneurial" universities (Etzkowitz *et al.*, 2000), and research laboratories and networks of supportive organisations; all factors identified by Parkinson *et al.* (2004). They combine to create a local entrepreneurial culture. The county also has good connectivity, including excellent connections to international airports, it has an attractive environment and a good quality of life, and the recent growth is based on diversity rather than focus on a single sector. OEO encapsulates this as the notion of a "triple-helix" of a small number of key elements which have underpinned the virtuous circle of activity (Figure 3.1). Parkinson *et al.* (2004) also highlight the strategic decision-making capacity of an area: the significance of networks and relationships between key players in the public and private sectors, the importance of crucial politicians in shaping strategies or influencing key programmes, and the significance of allies in national government. This report argues that the Oxfordshire cluster has not followed this virtuous path exactly. This theme is discussed later in this chapter.

Table 3.1. **Number of high-tech companies in Oxfordshire, analysed by sector, end of 2001**

Sector	Number of companies	As % of all high-tech companies
Manufacturing		
Publishing – specialist electronic only	9	0.6
Biotech, pharmaceuticals and medical diagnostics	73	5.2
Computer equipment	23	1.6
Electrical equipment (Butchart categories)	14	1.0
Electronic and telecoms equipment	46	3.3
Instruments, medical and optical equipment	112	7.9
Motorsport and automotive engineering/design	24	1.7
Aerospace and related services	12	0.9
Other manufacturing	70	5.0
Services		
Telecommunications	30	2.1
Software, web/Internet and other computer services	635	45.1
Other R&D activities (not included above)	44	3.1
Technical consultancy and technical testing	317	22.5
Other/not classified	8	–
Total: All sectors	**1 417**	**100.0**

Source: OEO 2003.

Table 3.2. **Sectoral breakdown of high-tech employment in Oxfordshire, end of 2001**

Sector	Number of employees, end of 2001	As % of all high-tech companies
Manufacturing		
Publishing – specialist electronic only	93	0.3
Biotech, pharmaceuticals and medical diagnostics	3 257	8.9
Computer equipment	1 825	5.0
Electrical equipment (Butchart categories)	657	1.8
Electronic and telecoms equipment	1 550	4.2
Instruments, medical and optical equipment	5 026	13.7
Motorsport and automotive engineering/design	2 503	6.8
Aerospace and related services	840	2.3
Other manufacturing	1 498	4.1
Services		
Telecommunications	2 335	6.4
Software, web/Internet and other computer services	7 899	21.6
Other R&D activities (not included above)	5 907	16.1
Technical consultancy and technical testing	3 257	8.9
Other/not classified	35	–
Total: All sectors	**36 682**	**100.0**

Note: Percentages rounded.
Source: OEO, 2003.

Figure 3.1. **The Oxfordshire virtuous circle of activity**

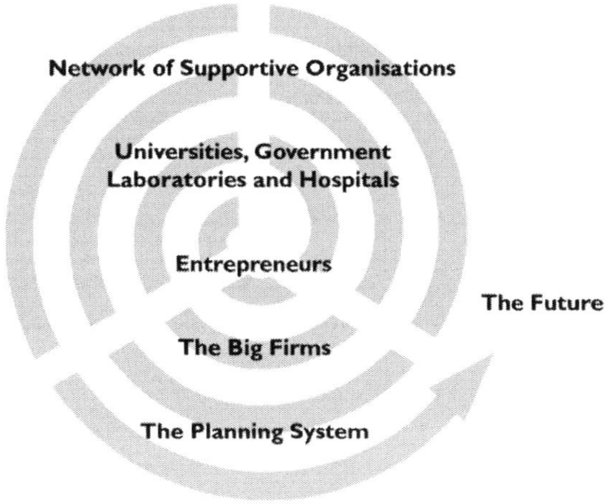

Source: Glasson *et al.* 2006.

Entrepreneurial individuals

At the centre of the virtuous circle are a set of entrepreneurial individuals and their companies, which have provided the big firm building blocks of the high-tech economy. Oxfordshire provides a contemporary example of what Richard Florida (2002) calls the "geography of talent" – a rich mix of entrepreneurs, philanthropists, local policy makers and professional services coming together to provide leadership, vision and examples of what can be done. In *Enterprising Oxford: the Growth of the Oxfordshire High-tech Economy* (2003) OEO drew on the information in a set of semi-structured biographies supplied by each of 22 individuals identified as having initiated change and played a part in developing Oxfordshire's enterprise culture. In addition to giving information on career highlights, and roles in national initiatives with impacts locally, each was asked to outline their own particular initiatives at the county level which had an impact on the Oxfordshire high-tech economy, and of their personal involvement in the high-tech economy. The individuals were divided loosely into "town" and "gown". The former comprise the entrepreneurs who have been responsible for leading the field in new technologies, plus those key individuals in the public, private and voluntary sector who have provided support in various ways. The "gown" group includes the star scientists, senior academics and administrators who have generated and processed inventions and innovations from the science-base. The distinction between the two groups is loose, and there are some important overlaps. The biographies show that although it was the 1980s when the high-tech economy first displayed rapid growth, and the 1990s when there was a professionalisation of the activity and of the HE-industry links, many of the entrepreneurial foundations had been laid much earlier in the 1960s and 1970s.

The outstanding example of Oxfordshire entrepreneurialism is provided by the work of Sir Martin and Lady Audrey Wood (Wood, 2001). The Woods epitomised the "can-do" mentality, and have been major players in the Oxfordshire high-tech economy since the 1950s. They were leaders in the field of cryogenics – using very low temperatures to develop high-tech superconducting magnets, which led to the development of Magnetic Resonance Imaging (MRI) and to medical instrumentation applications, most notably as body scanners. Their company, Oxford Instruments, generated many spin-offs, including Oxford Analytical Instruments, Oxford Magnetic Technology, and many others, giving rise to the phenomenon of Cryogenics Valley (Lawton-Smith, 1991). By 1999 Oxford Instruments four divisions: superconductivity, analytical, medical systems and magnet technology joint venture employed a total of 2 300. Other key entrepreneurs include Mike O'Regan, who along with Mike Fischer founded Research Machines (RM plc), which is the UK's leading supplier of information and communications technology to schools, universities and colleges; Paul Drayson, co-founder of Powderject who went on to become Minister of State for Defence in the House of Lords; Frank Williams, the driving force of the Formula 1 Williams motor racing team based at Grove in Oxfordshire (Henry *et al.*, 1996), and Jan Hruska, co-founder of Sophos, which offers computer anti-virus protection to university and business users.

The science-base

Underpinning such entrepreneurial activity is the science-base of Oxfordshire – its three universities, seven government laboratories and nine hospitals. Oxford now has two major science parks: The Oxford Science Park owned by Magdalen College and Oxford University Science Park at Begbroke. Oxford University's technology transfer company, Isis Innovation, is spinning out a company every two months and the University has over 30 millionaires. The Oxford Science Park has over 50 companies, mainly in computing and biotech. In 2001, Oxford University won a competition sponsored by US venture capital firm Cross Atlantic Capital Partners, for the being most enterprising UK University.

Oxford Brookes University also has a strong science-base, especially in biotech and automotive engineering (linked to the motorsports industry), and provides advanced training programmes in these areas. Oxfordshire also has the additional advantage of some of the most important government laboratories in the country, sitting astride the A34 – the United Kingdom Atomic Energy Authority (UKAEA) laboratories at Harwell and at Culham, the Central Laboratory of the Research Councils (CCLRC), the Rutherford Appleton

Laboratory (RAL), the Medical Research Council and several others. These laboratories in total, employ over 6 000 people and, over a period of 50 years, have provided an important home for science and invention. Audrey Wood (2001) highlights several of the Oxford University – RAL links, and the development of the flagship company of Oxford Instruments. A recent key boost to RAL is the government decision to site the Diamond Project, the next generation Synchrotron light source, there. This GBP 500 m facility is one of the largest research projects in the UK for 30 years. It becomes fully operational in 2006. Oxford is also a major centre for teaching and research hospitals (the John Radcliffe, Churchill, Nuffield), which play a vital role in the rich medical, biotechnological and pharmaceutical activity in the area.

Role of SMEs

New small and medium-sized enterprises (SMEs) have played a major role in driving the cluster. The cluster began to grow rapidly in the mid 1990s. Table 3.3 shows that the take off period of growth began in the period 1991-95, but accelerated from the middle of the 1990s. Table 3.4 shows that the rate of growth in Oxfordshire was the fastest in the UK during the 1990s. This is significantly faster than that of Berkshire, which has the highest absolute number of high-tech jobs, many of which are in multinational companies such as IT giant Microsoft and confectioner Mars.

Table 3.3. **Date of incorporation of high-tech companies in Oxfordshire, end of 2001**

Year company was first registered at companies house	Number of companies	% of total
2001-2002	29	3.1
1996-2000	357	38.1
1991-1995	233	24.8
1986-1990	139	14.8
1981-1985	92	9.8
1980 and Earlier	88	9.4
Year not known or company not registered	479	–
Total	**1 417**	**100.0**

Source: OEO 2003.

More recent official statistics for 2004 (Office for National Statistics) show a total of just over 3 500 businesses in Oxfordshire in high-tech and related sectors, with about 600 businesses in high-tech and related manufacturing sectors (mainly in publishing and instrument manufacture) and 2 900 businesses in high-tech services (mainly in software and other computer related activities and architecture, engineering and related technical consultancy). Overall employment in these businesses was about 37 000 in 2004, although this figure is

Table 3.4. **Rate of employment growth in high-tech sectors, 1991-2000, top 10 English counties (using Butchart definition)**

County	Growth	% growth
1 Oxfordshire	17 450	141.0
2 Berkshire	37 400	105.5
3 Cleveland	4 300	94.0
4 South Yorkshire	8 350	78.9
5 Nottinghamshire	8 000	56.7
6 Wiltshire	7 950	55.3
7 Lancashire	14 300	52.9
8 Cambridgeshire	9 650	52.7
9 Buckinghamshire	10 550	51.5
10 Surrey	16 700	49.7
South East Counties	104 400	51.1
All English Counties	274 850	28.5

Source: ONS, 2000 Annual Business Inquiry and 1991 Annual Employment Survey (OEO 2003).

affected by the reclassification of about 8 000-9 000 employees from R&D to higher education. If these employees are counted as part of the high-tech sector, then total employment is higher, at an estimated 45-46 000.

An interesting feature of the Oxfordshire high-tech cluster is that the overwhelming majority of the larger firms began as SMEs. These include Oxford Instruments (which employs some 2 000 people worldwide, half of whom are in Oxfordshire), Research Machines (1 000), and Bookham Technology (450). In 2002, the ten largest firms in the county employed 3 568: around 10 per cent of the total high-tech economy. Since then several of the high-tech SMEs have been acquired by foreign companies. Examples include Powderject by Chiron (later being acquired by Novartis) and Mirada by Siemens.

Spin-offs in Oxfordshire

A further role of SMEs in the Oxfordshire high-tech economy is that of raising the county's activity and profile as a location for science-based entrepreneurship, particularly in biotech and ICT. The county has a very high rate of university spin-offs, especially from Oxford University. The Oxford Economic Observatory (Lawton-Smith and Glasson, 2005) identified some 114 technology-based companies that have originated in the county's three universities (Oxford University, Oxford Brookes and Cranfield DCMT at Shrivenham) and seven research laboratories. These are firms which have been formed by a member of staff or a student, or on the basis of science/technology from the university. The study distinguishes between those spin-off companies that were founded to exploit the institution's intellectual property and those that were formed by staff or students in a technology field but not necessarily exploiting the university/laboratory's intellectual property.

The latter are termed "founder affiliates". On this basis, by March 2005 there were 64 spin-offs and 50 founder affiliates. The vast majority (c. 80 per cent) of both spin-offs and founder affiliate firms have been formed by Oxford University staff, and of those many have been established through Isis Innovation in the period since 1997.

The spin-offs date back to the early 1950s, when many of the early firms were instrument manufacturers. The two earliest, Littlemore Scientific Engineering 1954 and Oxford Instruments 1959 are still thriving. The late 1960s and 1970s saw greater diversity in the sectors in which the companies were formed, including studio recording equipment – Solid State Logic (1969), computing – Research Machines (1974) and lasers – Oxford Lasers (1977). The longevity of many of these firms is illustrated by the statistic that by 1987, some 35 firms (including consultancy companies not included here) already had their origins in Oxford University and nearly 40 per cent had been established by 1993. Recently the rate of formation has accelerated. Just over a third (40), were established between 1998 and 2004. Overall, nearly 90 per cent are still in existence. Even companies which have been dissolved stayed in business for many years. The survival rate is particularly high from Oxford University. Among the non-Oxford University spin-offs with founder affiliation are Psion and Oxford Applied Research, which were both formed by former Culham scientists, while Harwell Scientifics is a spin-off from Harwell.

Some the technology-based spin-offs have emerged as major UK companies, with a fifth of the companies having gone public: 12 are PLCs. After a gap of no IPOs for some years, three companies were launched on the Stock Exchange in 2004: VastOX, Evolutec and Physiomics. These joined other long established companies including Oxford Instruments and Research Machines PLC, all of which originated in Oxford University. In 2002, total turnover for the companies for which information is available was nearly GBP 1bn.

The OEO survey also includes analysis of the firms identified the connection between age of firm and employment generation, while other surveys have often only focused on the number of spin-offs (for example in annual surveys conducted by HEBI and UNICO in the UK and AUTM in North America). By 2002, the available data showed that spin-offs accounted for over 3 per cent of Oxfordshire's employment. The average firm size of 40 firms founded before 1994 rose from 140 in 1994 to 354 in 2001, with the rate of employment growth accelerating after firms had been established for ten years. Of the spin-offs created by the region's universities, most have stayed within Oxfordshire, so contributing to the extraordinary rate of growth in the county's high-tech cluster.

The OEO survey identified the greatest impact on employment to be concentrated in a few sectors, of which ICT is the largest single sector and biomedical is the largest composite sector accounting for 40 per cent of firms: 25 spin-offs are biotech and 16 are pharmaceuticals companies (Figure 3.2).

Figure 3.2. **Comparison of numbers of ICT and Biotech spin-offs**

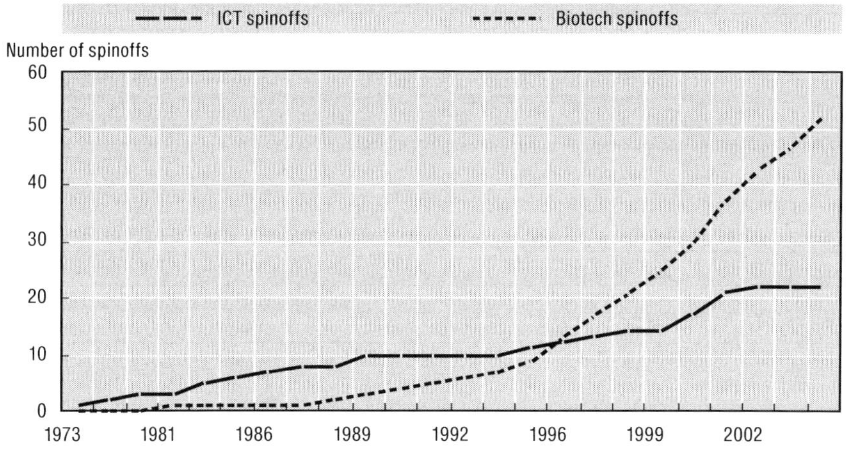

Source: Lawton-Smith and Glasson 2005.

Until 2001, the ICT group had more employees than the biomedical group, enjoying sharp employment growth between 1994 and 1996 when the growth slowed. Conversely, the biomedical sector has experienced a strong increase in the number of employees since 1997. This change is related to a correspondence with the rapid increase in the number of biomedical companies (Figure 3.3).

Impact of the cluster on entrepreneurship and employment

The growth of the cluster itself has had reinforcing effects on SME development and entrepreneurship, has contributed to a demand for governance systems, and has provided role models as above. The case study supports Berry and Glaeser's (2005), thesis that the tendency is for initially skilled places to become more skilled over time. They find that if skilled people are more likely to become entrepreneurs (who are relatively immobile) innovate in ways that employ other skilled people, then this creates an agglomeration economy where skilled people want to be around each other. This has a reinforcing effect of high skill locations and also has a series of spillover effects, including a demand for governance systems in the form of

Figure 3.3. **Employment data – comparison between ICT and biomedical group**

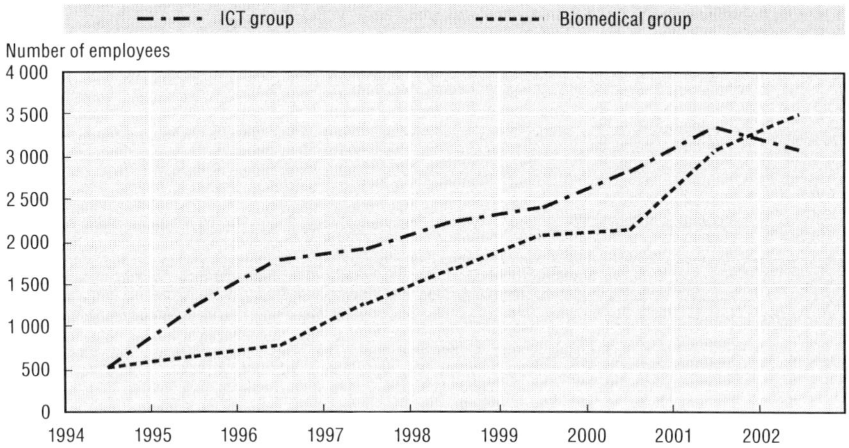

Source: Lawton-Smith and Glasson 2005.

training and financial systems etc., and through networks and coordination of knowledge through stocks and flows of people. In particular the cluster's growth has raised the level of skills in the county. Table 3.5 shows Oxfordshire is one of Britain's richest sources of intellectual capital having the second highest proportion of residents qualified to degree level of all 42 county areas in England and Wales, ranking seventh for the quantity of people so qualified. These skills provide the reservoir of talent from which future entrepreneurs will be drawn plus other kinds of creative people who service the firms – managers, lawyers, leaders of business communities and so on, who also represent the cluster in policy making circles at local, regional, national and international levels.

Table 3.5. **Educational attainment in Oxfordshire**

	NVQ level 4/5	NVQ level 4/5 Rank (of 376)	NVQ level 4/5	NVQ level 4/5 Rank (of 376)
South East	21.8	2 (of 9)	1 253 917	2 (of 9)
Oxfordshire	27.7	2 (of 42)	123 323	14 (of 42)
Cherwell	20.0	139	19 169	128
Oxford	36.9	12	38 301	38
South Oxfordshire	28.3	39	26 261	76
Vale of White Horse	28.2	41	23 576	87
West Oxfordshire	23.3	79	16 016	169
England	19.9		7 072 052	
England and Wales	19.8		7 432 962	

Source: Census data in Lawton-Smith and Waters, 2006.

Barriers to cluster development

Barriers to the development of the cluster have their history in lack of support from public policy makers, science park late development, a lack of co-ordination on labour market, weak regional development agencies, and the cluster's own success. In this section the focus is on the problems created by the rapid growth of the cluster as illustrated by a recent survey of the Oxfordshire biotech cluster.

The purpose of the Oxfordshire Bioscience Network (OBN) study was to examine the reality of the national and local business environment in Oxfordshire (see Lawton-Smith, 2005). Data presented here is from the Oxfordshire Bioscience Network 2001-02 postal survey. Replies were received from 75 firms out of an identified population of 100 firms employing 5 000 people.[1] The questionnaire was in three sections. The first was designed to identify the size and profile of the Oxfordshire biotech industry, in terms of growth rates, focus and funding cycles. This included address details, company profiles, specialist skills and technologies, key products and services, R&D and product focus, partnering opportunities sought and offered, year founded, number of employees, number of R&D staff and primary and secondary markets. The second, asked for further information about the firms including basic financial and employment data and issues relating to growth and sustainability of the biotech cluster in Oxfordshire – challenges, locational advantages, and education and training. The third asked for feedback regarding the type of sector network activity which the company felt added most value to the activity of the organisation. This included local networking events, the types of collaboration the firms already have, where they would like to develop new relationships, and sources of information and experience which are of most value to business development. In each of the three questions, respondents were asked to rate their answers on a 1-5 scale, where 1 is not significant and 5 is very significant). Scores were then added to give Table 3.6.

From this survey of the importance of the Oxfordshire biotech cluster development it appeared that university-industry links were not the main sources of information. Rather, local information sources were not the most important, conferences scored the highest, reflecting research practices which span both industry and academia. Other important sources were the Internet, published sources and trade fairs. Universities were ranked 9th in importance along with local sector networks, national trade associations, technology transfer departments and independent research organisations. Moreover, firms generally did not view proximity to Oxford University and the local research base as an important factor in the development of interactions, other than those of an informal nature. Networks including social interaction with friends, the second most important source of information, as well as

Table 3.6. **Locational advantages**

Rank	Advantage	Survey score
1	Proximity of the universities	236
2	Quality of life	221
3	Availability of staff	220
4	Proximity to like-minded companies	215
5	Proximity to airports	213
6	Availability of premises	206
7	Proximity to London	197
8	Availability of funding	178
9	Availability of specialist professional services	171
10	Access to investor networks	166
11	Access to sector networks	156
12	Access to mentors	154
13	Proximity to partner organisations	147
14	Proximity to market	124

Source: Oxfordshire Biosciences Survey 2002.

collaborators and ex-colleagues were also found to be significant. At the same time, competitors ranked as highly as friends. This is consistent with an earlier Oxford/Cambridge study which found that the majority of 50 firms surveyed in each region rated clients or customers as the most important external sources of innovation at the national and international scales (Lawton-Smith *et al.*, 2001).

The biggest drawback by far to an Oxfordshire location is the cost of housing (Table 3.7). This is followed by costs of living and costs and availability of laboratory space. High housing costs discourage people from moving to Oxfordshire and may encourage firms and people to leave the county. Salary costs are high as compensation for high living costs. Other elements of the physical infrastructure were important locational advantages and disadvantages. While Oxfordshire is a favourable location with regard to airports (ranked 5th) with access to London (with its frequent bus and train services ranked 7th), traffic and congestion costs were ranked as the 8th highest sector challenge, and therefore seen as a significant problem for firms in this industry.

Role of policy

At the early stages of the growth of the Oxfordshire high-tech cluster, national policy was delivered centrally. At the local level, there was a "statutory vacuum" (Lawton-Smith, 2003). A system of local governance rather than government appeared spontaneously during the 1980s which has been the main delivery of support for the high-tech cluster. It began with the formation of The Oxford Trust in 1985 when Dr. Martin Wood and his wife

Table 3.7. **Locational challenges**

Rank	Issue	Survey score
1	Cost of housing	247
2	Cost of living	228
3	Availability of laboratory space	225
4	Cost of office and laboratory space	221
5	Availability of scientists	215
6	Availability of appropriately qualified staff	213
7	Salary costs	211
8	Traffic and congestion	206
9	Inadequate public transport	202
10	Availability of technicians	200
11	Availability of managers	198
12	Availability of venture capital	197
13	Patent costs	187
14	Availability of business angel finance	178
15	Availability of office space	173
16	A lack of high-speed Internet access	168
17	Cost of specialist business advance	164
18	Cost of specialist equipment	164
19	Complex local planning and regulatory procedures	161
20	Current economic downturn	161
21	Availability of admin. staff	157
22	Availability of pre-clinical facilities	155
23	Availability of manufacturing facilities	155
24	Availability of bio-informaticians	152
25	Availability of specialist business advice	150
26	Availability of clinical trials facilities	132
27	Availability of distribution facilities	127

Source: Lawton-Smith (2005).

Audrey established The Oxford Trust. Its mission is "to encourage the study and application of science and technology" (Wood, 2001). Since its formation, in the absence of local policy frameworks, the Trust has taken responsibility for engaging with firms in the high-tech cluster and with research, business and public-sector organisations, speaking and acting on behalf of the high-tech sector locally and nationally. At the outset, the Trust recognised the need to network to provide innovation support. Its pioneering activities included managing the STEP Centre (incubator units) and facilitating networking through its innovation Forum, a series of seminars and workshops bringing together individuals and organisations on topics relating to business skills, developments in technology and future market opportunities, and running a schools programme. At the time of its formation it was the only place to which entrepreneurs could go for advice.

Oxford University was far from being entrepreneurial; the banks were largely unhelpful (Lawton-Smith, 1990); local government agencies were

CLUSTERS, INNOVATION AND ENTREPRENEURSHIP – ISBN 978-92-64-04442-5 – © OECD 2009

lacking in expertise; and contrary to Parkinson *et al.* (2004) local politicians were (and still are) uninterested and local authorities were under-resourced and unable to provide specialist advice. Oxfordshire County Council's low level of resources allocated to support economic development *per se* is a legacy of Conservative Party control. In line with national party policy of minimum state interference in the economy, the Authority's philosophy was to spend as little as possible. As a result, the County Council is the second-lowest spender behind Cornwall on many things including economic development. Party politics is also behind the City Council's lack of engagement with the high-tech sector and The Oxford Trust. The Labour Party, which has always controlled the City Council, has had as its priorities welfare, the low-wage economy and unemployment (Oxford City Council, 1997), rather than the high-tech economy. In this period, Oxford University and the government laboratories remained outside local networks. In the 1980s, Oxford University lagged behind Cambridge in the formation of institutions designed to support technology transfer. Until 1989, it had only a part-time industrial liaison officer; there was no Isis Innovation, Oxford University's technology transfer company, until 1988, while the equivalent in Cambridge had been set up in 1964. There was no science park until 1991 (1970 in Cambridge) because of greenbelt policies rather than a lack of proposals. The government laboratories were still effectively university-type research organisations, staffed by "risk-averse" scientists and engineers (Lawton-Smith, 1990).

During the 1990s, The Oxford Trust's activities expanded. In 1994, it moved to the Oxford Centre for Innovation, an incubator which it owns and manages; it established the Oxfordshire Investment Opportunity Network (OION)[2] in 1995 and the Oxfordshire BiotechNet consortium, a DTI initiative in 1997; and, through its subsidiary Oxford Innovation Ltd, it provided an increasing range of commercial activities including interim management, mentoring, advice and consultancy, services and premises. OION was one of the first business angel networks to be set up in the UK and links prospective investors with entrepreneurs seeking the early stage funding of between GBP 50 000 and GBP 1 million, which is often needed prior to seeking more substantial venture capital funding. OION has 80 investors and is now one of the most successful networks in Europe. Oxford Innovation Ltd now manages 10 business incubation centres in the county, with over 200 start-up companies.

In 2003, Gillian Pearson took over from Paul Bradstock, the founding Chief Executive of the Trust and in 2005 the Trust moved to a new and larger home in the centre of Oxford (Science Oxford). The Trust continues to assist the growth of technology businesses in Oxfordshire by:

- Highlighting their successes through an active website and published case studies.

- Providing practical information about starting or growing a new venture.
- Encouraging relationships between business, academia and government.
- Communicating the views of technology businesses to local and regional government.
- Analysing the growth of Oxfordshire's technology economy.
- Pioneering practical support activities in response to technology business needs.

All year round at Science Oxford there is a busy and varied What's-On programme with events, exhibitions and activities to engage business, schools and the public in discussion about science. These include drama and dance workshops and performances, sciart exhibitions, business networking lunches, improvised comedy and a regular series of Meet the Scientists as well as talks, debates and seminars on a wide range of science topics. It is also home to the Trust's interactive science gallery, Hands-On for young children, their teachers and parents.

Along with the continuing growth of the cluster, new networks have developed. These include the Oxfordshire BioScience Network launched in 1999 to improve networking and support to Oxfordshire bioscience enterprise, and "Venturefest" – "Oxford's International Fair for Entrepreneurs", held annually since 1999.[3] Each year "Venturefest" attracts in the region of 1 500 attendees to both the conference sessions and parallel exhibition. The profile of attendees has not changed markedly in the history of the event. Entrepreneurs continue to represent the highest number of attendees with the most common reason for attending being "seeking business partners". The second highest reason for attending is "seeking investment opportunities" further illustrating that the event achieves its aim of bringing ideas and funding together. Industries represented at "Venturefest" have not changed markedly either, although their percentage attendance might fluctuate slightly year on year. However, the top five industries are usually: Biotechnology, Education, Financial Services, Internet, Legal, Science/R&D and Software.

At the same time, in response to governmental priorities and the efforts of its key scientists, important organisational and cultural changes occurred at Oxford University and in Oxfordshire's government laboratories. Oxford University conducted a "review of technology transfer arrangements" in 1994. It was highly critical of the university's technology transfer policy and practice, and led to a series of changes, the most significant of which was the appointment of Dr. Tim Cook to head Isis Innovation in 1997. Dr. Cook had previously worked for Oxford Instruments and later became a serial entrepreneur, and thus brought much needed expertise to Oxford University's commercialisation company.

 CLUSTERS, INNOVATION AND ENTREPRENEURSHIP – ISBN 978-92-64-04442-5 – © OECD 2009

Changes were also taking place in the government laboratories. In 1996, part of the UKAEA was privatised leading to the formation of AEA Technology. The sites at Harwell in the Vale of White Horse and Culham in SODC remained the property of the UKAEA and have since been developed as a Science Centre (Culham) and a Technology Park (Harwell). The local consequences of the national agenda of valorisation were, first, an increasing level of interest from the government laboratories in local expertise in facilitating entrepreneurship and innovation, particularly with regard to the development of incubators.

In 2006, it was announced in the Budget, by the Chancellor of the Exchequer that there would be two science and innovation campuses – one in the South East and one in the North West. The South East one would be around the Harwell/RAL site owned by the CCLRC. The plan for this site is science + business and in middle, an interaction space. There are plans for conference centre, hotel, education and training centre to take account of activities which are on site and the Advanced Technology Regional Resource Centre. The resource centre is a joint SEEDA and European Social Fund project. CCLRC was given GBP 0.5 million in July 2006 to provide training under the arm of instrumentation technology specifically to SMEs in the South East Region. This includes electronics, advanced instrumentation and ICT.

Policy adaptations over time

In contrast to the Trust, the economic development activities of the Oxfordshire local authorities until recently have been quite limited. There have been few resources available, although since the late 1990s the County Council has sought to promote knowledge-based industries, innovation and lifelong learning as the underpinning of its emerging Economic Development Strategy (Oxfordshire County Council, 1999). In addition, there has been what Waters and Lawton-Smith (2002) refer to as a rescaling of organisational activity for high-tech development. The Oxfordshire Economic Partnership (OEP) and South East England Development Agency (SEEDA), established in 1998 and 1999 respectively, provide two new agencies of particular relevance for Oxfordshire. OEP seeks to unite non-statutory bodies, such as the Oxford Trust, with statutory bodies. It includes partners from local authorities, business, academia and the Oxford Trust, and worked jointly with the County Council to roll forward the Economic Development Strategy (OEP, 2001). One aspect of SEEDA's business strategic priorities is to "increase the rate of innovation and technology transfer by exploiting the world class research and development resources within the region" (SEEDA, 1999, 2002). With respect to cluster policy its main activities are Enterprise Hubs and regional networks (informal groupings of managers in particular industry groups who meet to discuss common problems and offer mutual support). Enterprise Hubs have a particular role in facilitating the transfer of technology and know-how from the region's universities and research institutes – with several of them being based on science campuses or science parks. Oxfordshire has two such hubs to date – for advanced materials and automotive engineering in the north of the county on the Oxford University Science Park at Begbroke, and for environmental, life and engineering sciences, in the south on the Harwell site.

As a consequence of these changes, the governance of the Oxfordshire high-tech economy is now more orchestrated as public policy at national, regional and local levels have focused on innovation-led development. The role of public policy in promoting clusters and overcoming barrier is a very recent phenomenon. At the national level, national interest in regional/local economic activity can be dated to the election of the Labour government in 1997. Key early dates were the publication of the 1998 Department and Trade and Industry (DTI) *Our Competitive Future* the agenda of which has the development of the knowledge economy; the establishment of the Regional Development Agencies (RDAs) in 1999 – which have responsibilities for cluster development and are increasingly given a greater role in stimulating and supporting innovation-led growth; the 1999 H-M Treasury report *Biotechnology Clusters: Report of a team led by Lord Sainsbury* (the Sainsbury Report); the 2001 DTI Clusters report (Business Clusters in the UK – a First Assessment). Since then there have been a number of other White Papers and Reports which have continued the theme of innovation, particularly the regional role of universities in stimulating economic development (see for example the 2003 Lambert Review of University-Business Interaction and the Innovation Report, "Competing in the global economy: the innovation challenge", which was published in December 2003 and also highlights regional innovation – in practice cluster strategies. A report for the DTI and the English RDAs, *The Practical Guide to Cluster Development,* was published in April 2004. Nationally, the government's flagship policy is tax relief on R&D and the Grant for R&D, which replaced the SMART programme but in a reduced form. This is now delivered by RDAs. Also new, and reflecting the recognition of the need to increase the supply of start-up and seed corn capital for SMEs, are the Enterprise Capital Funds (launched in July 2005) and the Regional Venture Capital Funds launched in 2006.

The RDA which covers the Oxfordshire area, the South East England Development Agency (SEEDA) The regional economic strategy for the South East 2006-16 – which like all previous regional economic strategies – has the vision that the South East will be a world class region achieving sustainable prosperity by 2016. Oxfordshire, although not specifically mentioned forms part of the Inner South East, thus a core part of the region's economy. By 2003, SEEDA's GBP 1.2 m Cluster Fund had seven cluster projects up and running with a further ten projects under development. In Oxfordshire this included funding for OXIT, a membership body which represents the IT industry in Oxfordshire.

However, one of the constant criticisms of SEEDA's initiatives is the low level of funding for initiatives. An example comes from the Oxfordshire Economic Partnership which represents key stakeholder in the Oxfordshire economy (*www.oep.org.uk*).[4]

> It is essential that Enterprise Hubs reflect the economy that they were developed out of, continually developing relationships with partners which enable the delivery of support services. The Hubs are only modestly funded and thus must work closely with a host of partners if they are to deliver their outputs. FE, HE and research institutions all play an important role in this, along with Business Links, local authorities, Learning and Skills Councils and Economic Partnerships and so on. These relationships need to be developed over time in the context of the local situation and will only work where trust and understanding have been built, which may take some time and clear evidence of delivery (OEP, 2004).

In 2006, SEEDA withdrew funding from The Oxfordshire BiotechNet incubator, which then closed. Oxford University's Department of Continuing Professional Development has approached SEEDA for funds to support local high-tech firms attend in-house training courses, but this has not been forthcoming.

Within Oxfordshire, Oxfordshire County Council has responsibilities for sustainable development that address the sometimes competing demands of economic development, protecting the environment, promoting social inclusion and reducing resource use. According to the County's Strategic Policy Manager, David Waller, while the issue of the green belt and making land available for economic development captures the headlines, the biggest driver of the economy is improving skills and providing effective business support. The South East Plan (SEERA, 2006) prioritises [ldquoe]smart"– high-value and resource efficient – growth, but that cannot be achieved without better skills than there are now in the economy. This brings about a renewed focus on the high-tech sector – where the main barrier to growth is skills availability rather than land availability. Recent changes in policy at the local/county level reflect an increased commitment to growing the economy as fast as possible without damaging the environment. Key initiatives include:

- **Local Area Agreement** – a relatively new and increasingly important structure initiated by Central Government that will increasingly provide a common framework of governance for funding to the county. It is led by the County Council and in the area of economic development and enterprise it targets improved adult skills, improved teaching of enterprise in schools and strengthening of the county's promotion of inward investment and international links.
- **Children and young peoples' plan** – local actors focusing together on improving outcomes for children and young people through the services

they provide as inputs. The Economy and Enterprise block of the Local Area Agreement has responsibilities for promoting a more enterprising culture in schools.

- **Oxfordshire Economic Partnership (OEP)** – an enhanced role for the OEP as the Local Area Agreement has led to the need to improve the translation of strategies developed by the partnership into effective delivery programmes. OEP acts as a network of networks bringing together public and private sector organisations involved in economic development to agree shared priorities and provide leadership to bring about the changes necessary to address them.

A further development is the Oxfordshire Data Observatory. What is missing in Oxfordshire is a publicly funded business database, as is maintained by Cambridgeshire County Council. The Oxfordshire Data Observatory is a database of databases. The intention is to provide access to information from a variety of information sources rather than develop in-house databases, and to commission research on specific topics.

Planning has been a key element in the way that the Oxfordshire high-tech cluster has developed. The Oxfordshire planning framework from the early days of structure planning was one of controlling rather than facilitating. Key building blocks in this framework were: extensive coverage of landscape designations, including the Oxford Green Belt; a country town strategy designed to spread housing and jobs to smaller towns outside Oxford; and a transport strategy designed to intercept commuting into the city – particularly via the Park and Ride policy. During the 1970s and 1980s, high-tech industry, with its desire to locate near to Oxford University, was generally discouraged, being seen as a threat to the Green Belt and the setting of Oxford. However, by the mid-1980s there was increasing concern with the potential implications of this policy. New companies in Central Oxfordshire, linked to research centres, had to move out to grow; and there was some national concern about their possible loss to the UK completely. Unemployment was also rising, and the County Council decided to introduce more flexibility into its employment policies.

Two key planning decisions were particularly important in supporting high-technology industry. The first was a 1987 Alteration to the Structure Plan, which added the new policy: "In Central Oxfordshire, provision will be made and proposals will normally be permitted for science-based industries concerned primarily with research and development which can show a special need to be located close to Oxford University or to other research facilities in Central Oxfordshire."

The Alteration also offered criteria against which further proposals for science park developments should be considered (*e.g.* sites should not be in the Green Belt, should be accessible by public transport and should be readily accessible from University science departments). There was a flood of applications, but only the Magdalen Science Park was approved in 1991, but this in itself was a major breakthrough for the County.

The second key planning decision relates to the 1987 revision of the Planning Use Classes Order (UCO) to include a new class – B1 Business Use – which allowed the combination of office space with light industrial operations for R&D – within one planning consent. In addition, the 1988 General Development Order (GDO) allowed for B2, General Industry Developments, to change to B1 without the need for planning permission. Together, these changes greatly facilitated the rapid growth of the Milton Park Business Park on the A34 near Didcot in the 1990s. By 1999, employment had increased to over 5 000, on an extensively landscaped Park covering nearly 250 acres. Amongst others, the Park is home to Research Machines, Evotec OAI Ltd., Psion Industrial, Oxford Semi-conductor and Bookham Technology. Other significant developments include new Business Parks near Oxford Airport (Elsevier Science HQ), and on the old Cowley motorworks sites.

Whilst these decisions were important catalysts in facilitating the high-tech growth, they also fuelled development pressures on the County, and the resultant (un)affordable housing, traffic congestion, and lack of technical skilled labour issues. For the future, the County needs smarter planning if it is to achieve sustainable development in all dimensions. Some encouraging features currently include a determination by local authorities to seek high proportions of affordable housing in new developments – for example 50 per cent plus 10 per cent for key workers for Oxford City; to consider innovative transport solutions – for example the proposed guided busway system; and to reconsider that most sacred of planning instruments, the Green Belt – with perhaps the introduction of green wedges, separating one or more "public transport/green corridor developments" under the current Oxfordshire Structure Plan Review (Oxfordshire County Council, 2004). More problematic is the current uncertainty in the planning system and the possibility of a vacuum in strategic planning between the SE Regional Spatial Strategy (RSS), and the District Local Development Frameworks. The initial planning for the SE RSS is wrestling with the coverage of sub-regional plans for the region, and Oxfordshire is presenting a particular challenge, lying as it does at the corner of the region, and with links across several regional boundaries. One of these, the Oxford-Cambridge Arc, is seen as a potential new axis, with Milton Keynes/Bedford in between benefiting from "high-technology overspill". Unfortunately, the necessary investment in rail infrastructure appears a long way off.

Future policy challenges

With respect to the involvement of the County and District councils in cluster development, the main debates are conducted through the SE Planning

Process and deal with issues that are Europe-wide. A particular issue is how should the spatial planning of economic development be organised? This is a green belt issue – whether land use constrains economic development and therefore should the strategy be to go flat out for economic growth? This raises the question of who bears the cost of such development. At the national level, the Chancellor of the Exchequer is interested in overall national GDP. Only three regions: the East of England, London and the South East contribute net receipts for the Treasury. For Oxfordshire, the cost of achieving national growth is borne locally at the cost of the quality of the environment – congestion, pollution, high housing costs etc., while the local gains are low levels of unemployment and a thriving economy. Most people are prosperous and have a good standard of living. The policy issue is how to match local and national policies. While nationally the aim is to ramp up economic development, local goals are different. There is no local public support for such development. Elected politicians have been voted in on policies which state that the green belt will be protected. Thus the environment has an economic as well as a social dimension. The autonomy of local planners in each of the districts further complicates matters. While it might be argued that the science parks at the public research laboratories at Culham and Chilton could be developed with little environmental damage, local planning decisions have restricted their development, especially in South Oxfordshire, the location of UKAEA Culham. In recent years the Vale of White Horse district, where Chilton is located, has been more supportive of science-led economic development. This made it possible for the science and innovation centre, announced by the Chancellor of the Exchequer, to be built on the RAL/Harwell site.

A more pervasive challenge is the potential threat from newly industrialising countries such as and India as well as from within Europe, for example from Germany which has better technician training than the UK. A potential outcome of such threats are that companies, particularly the increasing number of foreign-owned companies in the cluster, will relocate their activities to regions where there are greater numbers of highly skilled people, more beneficial taxation systems and better physical infrastructures.

Lessons for other clusters

The case of the Oxfordshire cluster highlights the significance of high-tech activity for economic growth in the 21st century. Oxfordshire has been highly successful in generating and exploiting knowledge, processes which according to the UK government's 1998 *Competitiveness* White Paper, are the foundations for a knowledge economy. While entrepreneurship is the cornerstone of cluster development, for policy makers, the cluster approach has to be seen as being about finding ways in which knowledge and innovation can be given a supportive environment (Malmberg and Power, 2006).

CLUSTERS, INNOVATION AND ENTREPRENEURSHIP – ISBN 978-92-64-04442-5 – © OECD 2009

The Oxfordshire cluster experience and analysis by Asheim *et al.* (2006) raise seven questions for the design of policies.

- How should entrepreneurship be encouraged?
- What can be done to solve the tensions between the needs of local firms and of planners to address the problems of pressures on the environment?
- How can the future growth of a cluster be secured through developing local capacity to accommodate change, especially within the labour market?
- Is there capacity in the building stock to meet the demand for incubators and space for expansion?
- How can we address social inequality between different groups of workers caused by the increasing demand for high level of skills and less demand for lower skills?
- How can growth in a latent or potential cluster be supported and existing clusters sustained?
- How can organisations established to support cluster development maintain vertical linkages and avoid being bureaucratic? How can the many initiatives be effectively co-ordinated and diverse interests appropriately balanced?

The implications for policy are first that mechanisms should be sought to promote high-tech activity. This is not only by finding ways to encourage the highly skilled to start new firms but also to help build a pool of talent that is needed to grow those firms – managers, financial advisors, patent agents and so on. This also includes mechanisms to increase the supply of start-up and early stage financing, vital in supporting entrepreneurship which is key to economic development (Lipper and Sommer, 2002). One of Oxfordshire's great strengths is its considerable number of business angels but more needs to be done to solve the "equity gap" in funding for start-up and early stage development phases.

Moreover, entrepreneurship could be encouraged by national government which could do more to help small firms – for example it could adopt the US Small Business Innovation Research Program (SBIR). The SBIR is targeted at the entrepreneurial sector. A specific percentage of federal R&D funds is reserved for small business. Thus SBIR protects the small business and enables it to compete on the same level as larger businesses. SBIR funds the critical start-up and development stages and it encourages the commercialisation of the technology, product, or service (*www.sba.gov/sbir*).

The second main challenge for Oxfordshire is supporting rapid high-tech economic growth while at the same time maintaining the environmental qualities that are one of the county's major assets particularly to the local planning authorities. Negative restraint policies will not be adequate to achieve the "smart growth" needed to meet national objectives. Their legacy from the past has been increasing traffic congestion and reduced agglomeration economies as a result of spreading development around the country towns (Witney, Banbury, Bicester and Didcot) rather than concentrating in larger centres on public transport routes. The policy implication for other places faced with issues of growth and environmental protection is that a clear and proactive policy vision for the next 15-20 years will be needed, and the management of change towards this vision will need to operate at least at the same pace as the private sector economic changes in the county. In Oxfordshire, the revised County Structure Plan (Oxfordshire County Council, 2004) and the emerging Sub-Regional Strategy for Central Oxfordshire, in the context of the developing SE Regional Spatial Strategy (SEERA, 2004) include first steps in seeking to better manage the conflicting environmental pressures. This is an outcome of a dialogue between local, regional and national government. Thus it is essential for good communications to be maintained between all relevant tiers of government and the private sector, for example through Chambers of Commerce, and with local universities and public research laboratories.

The third point is particularly pertinent in the Oxfordshire case. Emerging economic constraints relate to shortages in middle-level/technician level skilled labour, and to still not enough investment in R&D when benchmarked against international best practice (SEEDA, 2001; EC, 2004). The sustainability of Oxfordshire's high-tech economy requires investment not only in new intellectual capital and qualified people. A key problem now and in the future is the further raising of the level of skills and the provision of appropriate training for the high-tech cluster's growing number of firms. Therefore it is essential to have high quality data on current trends and analyses of how that translates into forecasts for skill and property needs, as well as a policy framework which will be able to respond to those needs. Skills are now top of the Oxfordshire and regional agenda as it is recognised that the cluster's growth will not be sustained without an adequate supply of highly skilled – therefore well trained – people. This growing emphasis on skills is to be welcomed, but Oxfordshire still has some way to go before there is anything like a coherent local approach to skills development.

Fourth, the Oxfordshire cluster is well provided with incubators for the start-up phase of new company formation, but there is an increasing difficulty in finding premises in locations, convenient for suppliers and the universities, to house the expansion phase of high-tech activity. The policy implication is that if firms are to make a major contribution to the local, regional and national export base then the physical infrastructure for the growth phase of development should be appropriately planned for and designed. This may mean some modifications to existing land use priorities. The Oxfordshire high-tech cluster faces an increasing "infrastructure deficit" in supporting the expansion of high-tech activity. Problems are especially associated with the transition of its high-tech firms to more mature stages. Moreover, expansion

 CLUSTERS, INNOVATION AND ENTREPRENEURSHIP – ISBN 978-92-64-04442-5 – © OECD 2009

is limited by policies which restrict the in-migration of large scale production units. These are prohibited under current planning rules because of the resulting pressure on housing and transport.

Fifth, public sector policy makers are also increasingly concerned about issues of social sustainability. High-tech development can raise issues of social cohesion, by virtue of the nature and type of workforce required. At the top, very high levels of formal qualifications in the emerging sectors are needed in order to plan and design the transfer of scientific knowledge into commercial products and services. Lower down, degree level skills are still required in order to organise and produce these new outputs. But what contributions can workers without such qualifications make to the changing economy; will they be excluded from the rewards of change? There is also the associated issue of whether workers at several levels can afford to live in the expensive housing of a "hotspot" high-tech economy. Policy must therefore address issues of social inequality which appear to be a common outcome of high-tech led economic development. Compensation through local and national tax spend may be necessary, combined with a programme targeted at retraining workers who lack relevant skills. Hence it is necessary for high-tech firms, schools, colleges and universities to work together in order to come up with possible solutions.

Sixth, in order for high-tech clusters to be sustained, policies needed to be developed on the basis of intelligence gathering of current and future activity so that there can be adequate planning for growth. An ongoing challenge is to identify latent or emerging clusters and provide a supportive environment for their future growth. In Oxfordshire, environmental technologies have been identified as a sector that is likely to expand. This is a reflection of national and international policy agendas relating to climate change and other environmental issues such as recycling and land management. Other developing clusters are around nanotechnologies and new materials. For those and other clusters to develop quickly and to be sustained, planning needs to be in place to anticipate the future demand for business units, homes, skills and transport infrastructure. A considerable challenge is in orchestrating a coherent response from the range of bodies responsible for decisions in these areas. Moreover, policy makers should be aware that when even high-tech industries mature, there can be a loss of jobs when companies are merged or acquired. Even though the biotech sector is still growing in Oxfordshire, merger and acquisition of many of its leading firms has meant some loss of jobs and reorientation of activity.

Seventh, the Oxfordshire cluster has benefited from an extensive set of networks and initiatives. A current study by OEO has identified five network hosts (organisations which hold networking events) and 62 business to business networks. Of the latter, five are related to raising finance.

Partnerships should be built between firms directly and collectively in the form of initiatives that promote the conditions for a supportive environment. Networks and networking are a key element in developing a coherent strategy. Evidence from other studies suggests that networks have at least three beneficial effects:

1. Networked businesses are likely to be more successful than non-networked business. Membership is linked to small business survival and, though networking with competitors, results in firms having a greater knowledge of their own strengths and weaknesses and a greater knowledge of the industry (Besser *et al.*, 2006). Mutual support networks enable their members to become more competitive through improved marketing and innovation, sharing of best practice, and access to current research, collective action and infrastructures.
2. Networked firms are more innovative. A review of networking and innovation in the UK by Pittaway *et al.* (2004) confirmed that networks and networking amongst firms plays a pivotal role in innovation and that this has become more relevant as technologies become more complex. The use of networks is crucially important during venture formation and for small growing firms.
3. Not only do individual firms benefit directly, networks act as "open gates" bringing in new ideas and practices to the local economy as a whole (Eradin and Armatli-Koroglu, 2005).

As the Oxfordshire high-tech cluster has grown, an array of supportive organisations has grown up. Such partnerships and networking have been vital ingredients in the development of the cluster, and key individuals have helped to change underlying mindsets and the environment that shape the nature of economic growth. In Oxfordshire they have originated more from charitable organisations and the private sector than local government while central government and more recently the regional development agencies have played a vital role in "pump-priming" networking activity.

It is likely that new organisations will develop networks and in line with its regional economic development strategy, the regional development agency, SEEDA, will increase its networking activities and the supply of start-up funding through the nationally supported Regional Venture Fund initiative. On the other hand, it has withdrawn funding from two which in the case of one, the Oxfordshire BiotechNet, this had the consequence that the networks and its incubator had to close. An issue for this and other policy making bodies is for how long initiatives such as networks and incubators should be underpinned by public money? Should the funding be permanent which might have the effect of making the initiative inefficient, or because of the "public good" nature of the programme, should it be sustained over a longer period of time in order to maximise the benefit to the community at which it is targeted?

An example of an organisation which may establish networks is the Oxford to Cambridge Arc, not necessarily in established technologies such as biotech and biopharmaceuticals where networks are well developed and linked across the Arc through, for example, academic networks. The potential for emerging and disruptive technologies is where the Arc can help to create a critical mass – which is not there at either end. The Arc could play a significant role in bringing networks together. The policy implication is that scope for collaboration between clusters should be explored, but such activity might need to be underpinned by public agencies.

Acknowledgements

This review draws on interviews with:

Professor Peter J. Dobson, Academic Director, Oxford University Begbroke Science Park.

Sam Foster, PR Department, CCLRC.

Rebecca Lingwood, Director, Continuing Professional Development, Oxford University.

Joe Barclay, Regional Liaison Director University of Oxford, Business Liaison Unit.

Ed Metcalfe, Head, Science Technology, Entrepreneurship and Management Learning and Skills, SEEDA.

Gillian Pearson, CE, The Oxford Trust.

Fiona Reid, Director, Oxford Science Enterprise Centre and Entrepreneurship Said, University of Oxford, Said Business School.

Dave Waller, Manager of the Strategic Policy and Economic Development Unit, Oxfordshire County Council.

Notes

1. Using a more restricted definition, OEO (2003) identifies 73 firms.
2. *www.oion.co.uk.*
3. *www.venturefest.com.*
4. Economic partnerships are statutory requirements.

References

Asheim, B., P. Cooke and R. Martin (2006), "The Cluster Concept in Regional Analysis and Policy" in B Asheim, P. Cooke and R Martin (eds.), *Clusters and Regional Development: Critical Reflections and Explorations,* Routledge, London.

Berry, C. R. and E. L. Glaeser (2005), "The Divergence of Human Capital Levels across Cities", *Harvard Institute of Economic Research Discussion Paper 2091.*

Besser, T. L., N. Miller and R. Perkins (2006), "For the Greater Good: Business Networks and Business Social Responsibility to Communities", *Entrepreneurship and Regional Development,* Vol. 18, pp. 321-339.

Butchart, R. L. (1987), "A New Definition of High Technology Industries", *Economic Review,* 400, pp. 82-88.

Chadwick, A., J. Glasson, H. Lawton-Smith, J. Simmie and G. Clark (2003), *Enterprising Oxford: The Anatomy of the Oxfordshire High-tech Economy*, Oxfordshire Economic Observatory, Oxford.

Department of Trade and Industry (DTI) (1998), *Our Competitive Future: Building the Knowledge Driven Economy,* DTI, London.

DTI (2001), *Regional Competitiveness Indicators, Annex 1 – Definitions,* September 2001.

Eradin, A. and B. Armatli-Koroglu (2005), "Innovation, Networking and New Industrial Clusters: the Characteristics of Networks and Local Innovation Capabilities in the Turkish Industrial Clusters", *Entrepreneurship and Regional Development,* Vol. 17, pp. 237-266.

Etzkowitz, H., A. Webster, C. Gebhardt and B. R. Cantisano Terra (2000), "The Future of the University and the University of the Future: Evolution from Ivory Tower to Entrepreneurial Paradigm", *Research Policy,* Vol. 29, No. 2, pp. 313-330.

European Commission (EC) (2004), *Third Report on Economic and Social Cohesion.* EC, Luxembourg.

Florida, R. (2002), *The Rise of the Creative Class*, Basic Books, New York.

Garnsey, E. and H. Lawton-Smith (1998), "Proximity and Complexity in the Emergence of High Technology Industry: The Oxbridge Comparison", *Geoforum,* Vol. 29, No. 4, pp. 433-450.

Glasson, J., A. Chadwick and H. Lawton-Smith (2006), "The Growth of Oxfordshire's High-Tech Economy", *European Planning Studies,* Vol. 14, No. 4, pp. 503-524.

Henry, N., S. Pinch and S. Russell (1996), "In Pole Position? Untraded Interdependencies, New Industrial Spaces and the British Motor Sport Industry", *Area,* Vol. 28, No. 1, pp. 25-36.

HM Treasury (2003), *Lambert Review of Business-University Collaboration* (The Lambert Report), *www.hm-treasury.gov.uk/consultations_and_legislation/lambert/consult_lambert_index.cfm.*

Lawton-Smith (1990), *The Location and Development of Advanced Technology in Oxfordshire in the Context of the Research Environment,* unpublished DPhil thesis University of Oxford.

Lawton-Smith, H. (1991), "The Role of Incubators in Local Industrial Development: The Cyrogenics Industry in Oxfordshire", *Entrepreneurship and Regional Development,* Vol. 3, pp. 175-194.

 CLUSTERS, INNOVATION AND ENTREPRENEURSHIP – ISBN 978-92-64-04442-5 – © OECD 2009

Lawton-Smith, H., J. Glasson, J. Simmie, A. Chadwick and G. Clark (2003), *Enterprising Oxford: The Growth of the Oxfordshire High-Tech Economy*, Oxfordshire Economic Observatory, Oxford.

Lawton-Smith, H., D. Keeble, C. Lawson, B. Moore and F. Wilkinson (2001), "Technology Transfer in Two Elite Locations: the Oxford and Cambridge Regions", *Tijdschrift Voor Economiche en Sociale Geographie*, Vol. 1, pp. 88-99.

Lawton-Smith, H. (2003), "Local Innovation Assemblages and Institutional Capacity in Local High-Tech Economic Development: The Case of Oxfordshire", *Urban Studies*, Vol. 40, No. 7, pp. 1353-1369.

Lawton-Smith, H. (2005), "The Biotechnology Industry in Oxfordshire: Enterprise and Innovation", *European Planning Studies*, Vol. 12, No. 7, pp. 985-1002.

Lawton-Smith, H. and J. Glasson (2005), *Public Research High-tech Spin-offs: Measuring Performance and Growth in Oxfordshire*, Oxfordshire Economic Observatory, Oxford.

Lipper, G. and B. Sommer (2002), "Encouraging Angel Capital: What the US States are Doing", *Venture Capital: An International Journal of Entrepreneurial Finance*, Vol. 4, No. 4, pp. 357-362.

Malmberg, A. and D. Power (2006), "A Severe Case of a Conceptual Headache" in B. Asheim, P. Cooke and R. Martin (eds.), *Clusters and Regional Development: Critical Reflections and Explorations*, Routledge, London.

OECD (1997), "Revision of the High Technology Sector and Product Classification", STI Working Papers 1997/2, OECD, Paris.

Oxfordshire County Council (2002), *Review of the Oxfordshire Structure Plan – Issues Paper*, Oxfordshire County Council, Oxford.

Oxfordshire Economic Partnership (2001), *Oxfordshire Economic Development Strategy 2001-2005*, December 2001.

Oxfordshire Economic Partnership (2004), *The Impact of SEEDA on the regional economy through Enterprise Hubs and Gateways*, A Response to the South East England Regional Assembly March 2004, *www.southeast-ra.gov.uk/meetings/select/enterprise_hubs/160404/oxfordshire_economic_partnership.pdf.*

Oxfordshire Economic Observatory (2007), *Enterprising Oxford: The Oxfordshire Model*, Oxfordshire Economic Observatory, Oxford.

Parkinson, M., M. Hutchins, J. Simmie, G. Glark and H. Verdonk (2004), *Urban Renaissance Characteristics of EU Non-capital Cities*, ODPM, London.

Pittaway, L., M. Robertson, K. Munir, D. Denyer and A. Neeley (2004), *Networking and Innovation in the UK: A Systematic Review of the Literature*, Aim Research, *www.aimresearch.org/publications/networking.pdf.*

Sainsbury Report (1999), *Biotechnology Clusters: Report of a Team led by Lord Sainsbury, Minister for Science*, Department of Trade and Industry, London.

SEEDA (1999), Regional *Economic Strategy: Building a World Class Region*. SEEDA, Guildford.

SEEDA (2002), *An Economic Profile of the South East*, SEEDA, Guildford.

SEEDA/Robert Huggins Associates (2001), *Global Index of Regional Knowledge Economies: Benchmarking SE England*, SEEDA, Guildford.

SEERA (2004), "Integrated Regional Framework", *www.southeast-ra.gov.uk/our_work/planning/sus_dev/irf.html.*

Waters, R. and H. Lawton-Smith (2002), "Regional Development Agencies and Local Economic Development: Scale and Competitiveness in High Technology Oxfordshire and Cambridgeshire", *European Planning Studies*, Vol. 10, No. 5, pp. 633-649.

Wood, A. (2001), *Magnetic Venture: the Story of Oxford Instruments*, OUP, Oxford.

CLUSTERS, INNOVATION AND ENTREPRENEURSHIP – ISBN 978-92-64-04442-5 – © OECD 2009

ISBN 978-92-64-04442-5
Clusters, Innovation and Entrepreneurship

Chapter 4

The Biotechnology Cluster of Vienna, Austria

by
Franz Tödtling and Michaela Trippl,
Institute for Regional Development and Environment,
Vienna University of Economics and Business Administration

This chapter examines the development of the Vienna biotechnology cluster and the role of policy actions in its evolution. This contribution challenges the prevailing view that the development of high-technology industries is always a spontaneous phenomenon. It will be argued that in the case of regions such as Vienna which have weak potentials for high-technology industries, the development of biotechnology clusters can be promoted with proactive policy efforts to create a favourable environment for high-technology activities. Furthermore, it will be shown that, besides the leading role played by large firms, well co-ordinated policy approaches between the national and regional level are important for developing a biotechnology industry in a region.

Introduction

Biotechnology is acknowledged to be a key future growth area of the emerging knowledge-based economy (Cooke, 2002; OECD, 2004), exhibiting a strong potential for dynamism and innovativeness. Being a technology of generic character, biotechnology has strong cross-sectoral effects, brought about by its application to many sectors including medicine, agriculture, food, etc. Biotechnology can be regarded as a strongly science-based industry, drawing on an analytical knowledge-base (Asheim and Gertler, 2005). This implies that it is dominated by the use of abstract and codified knowledge (Fuchs and Krauss, 2003). At the same time, complementary tacit knowledge is crucial for innovation processes in biotechnology (Zucker *et al.*, 1998; Oliver, 2004), underpinning the industry's strong spatial concentration tendencies in local clusters (Feldman, 2000; Lawton-Smith and Bagchi-Sen, 2004).

The key aim of this chapter is to get a better understanding of the role of policy actions for the development of biotechnology clusters. Based on an in-depth study of the Vienna biotechnology agglomeration we challenge the prevailing view in the literature that the birth and early evolution of such clusters are always spontaneous phenomena. It will be shown that in the case of regions with weak potentials for high-technology industries the rise of biotechnology clusters is more dependent on proactive policy efforts to create a favourable environment for high-technology activities (see also Trippl and Tödtling, 2007). Furthermore, it will be argued that well co-ordinated policy approaches between the national and regional level are important for developing a biotechnology industry in a particular region.

This chapter has been written on the basis of available policy documents, literature and data as well as of research results collected in the context of two projects on knowledge-based industries in Austria.[1] In the Vienna biotechnology cluster, 43 qualitative interviews have been taken with representatives from firms, knowledge providers and the regional policy and supporting systems, using semi-standardised questionnaires. Furthermore, seven representatives from the national policy and supporting system have been interviewed.

The remainder of this chapter is organised as follows: the chapter begins with an overview on the structuring of the biotechnology cluster and its evolution during the past ten years. This is followed by an analysis of the drivers of the competitiveness of the cluster and an examination of the role of SMEs for its development. The impact of the cluster on SME development and

CLUSTERS, INNOVATION AND ENTREPRENEURSHIP – ISBN 978-92-64-04442-5 – © OECD 2009

regional innovation is discussed. Then, key barriers curtailing the dynamism of the cluster and challenging its future evolution and dynamism are identified. This is followed by an analysis of the role of public policy at national and regional levels in promoting the Vienna biotechnology cluster. In the remaining sections recent policy innovations and future policy challenges for the cluster are highlighted. Finally, some conclusions are made by elaborating on general lessons that could be drawn from the experiences of the Vienna biotechnology cluster.

Nature and evolution of the cluster

Vienna's biotechnology cluster shares some features with the Austrian biotech sector: It is specialised on medical ("red") biotechnology (Baier *et al.*, 2000; Oosterwijk *et al.*, 2003). Like Austria it has to be regarded as a latecomer in the commercialisation of biotechnology. Although there is a good scientific base the commercial exploitation of research results is underdeveloped (Reiss *et al.*, 2003). This is mainly due to weak incentives and conditions for commercialising research predominating in the past and a lack of tradition and culture for high-risk taking.

The Austrian biotechnology sector is characterised by a strong tendency toward spatial concentration. More than 65 per cent of all biotech related firms are located in the Vienna region. Smaller clusters of biotechnological activities can be found in the provinces of Styria, Lower Austria and Tyrol (see Table 4.1). In the following section the main structuring of the Vienna biotech cluster will be analysed.

Table 4.1. **Biotechnology related companies in Austrian provinces**

Region	Number of firms	Proportion of firms (%)
Vienna[1]	74	66
Styria	10	9
Lower Austria	10	9
Tyrol	9	8
Upper Austria	4	4
Salzburg	4	4
Vorarlberg	1	1
Burgenland	0	0
Carinthia	0	0
Total	**112**	**100**

1. Data for Vienna for the year 2006, data for other provinces for the year 2005.
Source: BIT and LISA (2004), complemented by our own inquiry.

Nature of the cluster

The Vienna region is Austria's most important centre for medical biotechnology, hosting more than 70 biotechnology related firms. The structure of the cluster is characterised by the existence of six subsidiaries of foreign big pharmaceutical companies, 33 small dedicated biotechnology companies, and several specialised and other suppliers (see Table 4.2):

- There is a long time presence of subsidiaries of multinational companies in the region which had been attracted in the post-war period by the strong research base and the easy recruitment of highly skilled scientists. The key players are Boehringer Ingelheim, Novartis, and Baxter. Boehringer Ingelheim Austria includes the company's centre for cancer research, one of its two centres of competence in biopharmaceutical production and its basic research subsidiary IMP. Novartis is the nation's largest pharmaceutical producer with a total of more than 3 000 workers. Baxter Austria is the company's most important research operation outside the US. Another key actor is Eli Lilly which mainly carries out clinical research projects in the area. Furthermore, there are about 60 sales and distribution firms located in Vienna. Among these are subsidiaries of Amgen, Aventis, Behring and Schering that all see the region as a sales and distribution centre for the Eastern European market.
- The Vienna biotech cluster, furthermore, hosts 33 dedicated biotech companies. Examples include Intercell (vaccines against oncological and infectious diseases), Igeneon (oncology) which has recently been acquired by the US biopharmaceutical company Aphton, Austrianova (oncology, gene therapy) or Green Hills Biotechnology (oncology). About 40 per cent of the dedicated biotech firms were founded within the past five years and many of them employ less than ten workers. There are about 20 specialised suppliers operating in the area. This segment mainly consists of producers of research agents (Bender Med Systems, Nano-S), bioinformatics providers (Emergentec, Insilico) and firms performing clinical trials services. Finally, there are ten suppliers offering laboratory products and equipment.
- Venture capital firms and business angels are a missing ingredient in Austria (Senker, 2004) and in the cluster. There are few such firms, *e.g.* Horizonte Venture Management (Technopolis, 2006). The main reason for this is the bank-dominated landscape with a preference for traditional credit instruments and a widespread adversity to risk taking. Consequently, successful companies like Intercell or Igeneon had to attract external financing from international venture capitalists and funds.

An analysis of the fields of activity of the cluster firms (Table 4.2) reveals that the multinational companies and the biotech firms are mainly specialised in the segment therapeutics. Moreover, there is a strong presence of producers of laboratory products, diagnostics and consulting in the Vienna biotechnology cluster. Bioinformatics, in contrast, only plays a minor role.

CLUSTERS, INNOVATION AND ENTREPRENEURSHIP – ISBN 978-92-64-04442-5 – © OECD 2009

Table 4.2. **Classification of biotechnology related firms in Vienna**

Classification	Number of firms	%	Fields of activity	Number of firms	%
Big pharma	6	8			
Biotech firms	33	45			
			Therapeutics	26	35
			Diagnostics	9	12
			Clinical research	4	5
Specialised suppliers	22	30			
			Clinical research	3	4
			Bioinformatics	3	4
			Consulting	7	10
			Research agents	4	5
			Other activities	5	7
Other suppliers	10	14			
			Laboratory products	10	14
Other firms	3	4	Other activities	3	4
TOTAL	**74**	**100**		**74**	**100**
Sales and distribution Firms	61				

Source: Tödtling *et al.* (2006, p. 17).

Exact data on the number of employees are not available. A recent study (Technopolis, 2006) estimates that the three Big Pharma companies Boehringer Ingelheim, Novartis and Baxter employ about 3 700 people and that the biotech firms employ about 1 500 workers.[2]

Evolution of the cluster

In the past decade, Vienna biotechnology has shown a relatively good growth performance. The evolution of the local industry in this period was mainly characterised by new firm creations and the emergence of more differentiated and specialised cluster structures. In the last ten years, the number of cluster firms has been growing (Figure 4.1). In particular since the end of the 1990s there has been an intensification of new firm formation in the Vienna biotechnology cluster. Since 2003, however, the rate of new business establishment has slowed down. A more detailed analysis shows that the dynamic development of the cluster since the mid 1990s mainly rests on the setting up of new ventures in the fields of therapeutics and specialised supply (Figure 4.2). Consequently, in the past few years a specialisation on more high-value activities has taken place.

The specialisation process of the cluster structures is also reflected in the research landscape. Recently, a strengthening and further differentiation of the cluster's research and education capabilities set in, brought about by the establishment of new specialised institutes (see next section).

Figure 4.1. **New firm formation in the Vienna biotechnology cluster**

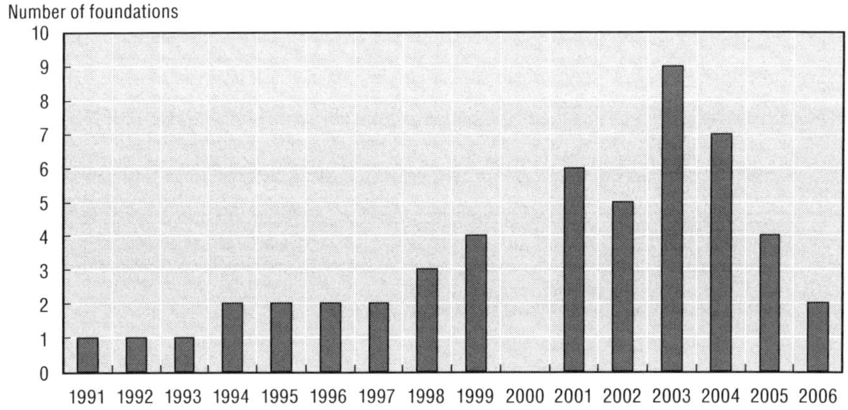

Source: Tödtling *et al.* (2006, p. 19).

Figure 4.2. **New firm formation in different fields of activity**

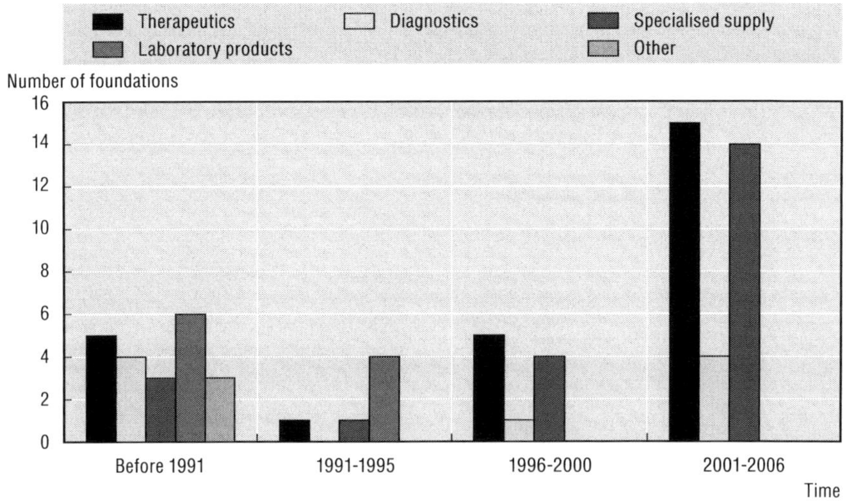

Source: Tödtling *et al.* (2006, p. 20).

Success factors

In international comparison the Vienna biotechnology cluster is rather young, small and in a catching-up position. Its emergence, however, dates back a couple of decades. As already mentioned above, Austria and especially its capital Vienna have an excellent tradition in medicine and biomedicine (Oosterwijk *et al.*, 2003). Already in the 1950s a first successful spin-off company from the Vienna University, Immuno, entered the scene.[3] We might identify three subsequent stages of development:

- The period between the 1950s and the 1980s was shaped by the early arrival of Big Pharma. The strong research base and the easy recruitment of highly skilled scientists had attracted international pharmaceutical companies. Boehringer Ingelheim settled down in 1949, Novartis arrived in 1969 and Baxter opened its doors in 1983.
- In the next phase from the 1980s to the midst 1990s there were first steps to enter into the new technology. In this period, two academic spin-offs (Technoclone and Nanosearch Membrane) have been founded. They remained, however, individual cases as commercialisation of biotechnology was still in its infancy in the Vienna region. More important for the cluster's development was the foundation of the Institute of Molecular Pathology (IMP) in 1988 with substantial support from both national and regional policy makers. The IMP is an internationally renowned basic biomedical research centre sponsored by Boehringer Ingelheim. IMP's arrival gave rise to the foundation of the new campus Vienna Biocenter housing space for the IMP itself and several university institutes with a focus on molecular biology.[4]
- Since the end of the 1990s, in a third stage of development the cluster seems to gain momentum. What can be observed is a rise in commercialisation of biotechnological research by an intensification of academic spin-off activities. Moreover, policy makers and public institutions are also enforcing their efforts to promote the new industry.

The excellent scientific system and research capabilities have not only been crucial for the emergence of the Vienna biotechnology cluster, but are also the key driver of its contemporary competitiveness. The subsidiaries of the Big Pharma companies Boehringer Ingelheim, Novartis and Baxter all have established research centres in the region of Vienna (see above). The cluster's attractiveness for these business research activities is to be seen in its excellent scientific base comprising five universities, several hospitals and a range of other public and private research institutes. There is an estimate of about 3 800 researchers in the field of Life Sciences employed in the region (Technopolis, 2006). The annual growth rate of publications amounts to 11 per cent (see Table 4.3). More than 50 per cent of all scientific publications and 64 per cent of all patent applications could be found in the fields of therapeutics and diagnostics (Technopolis, 2006), revealing a strong specialisation in key future areas of biotechnology.

We may identify five different research nuclei in the cluster, underpinning its dynamic development and competitiveness (Figure 4.3).

The most central research core is the Campus Vienna Biocenter set up in 1992. It currently employs over 1 000 scientists from 40 nations. It

Table 4.3. **Publications in biotechnology per year**

	Number of publications
1995	202
1996	287
1997	259
1998	301
1999	321
2000	334
2001	390
2002	425
2003	427

Source: Technopolis (2006).

encompasses the Vienna Biocenter (VBC) containing eight departments of the University of Vienna and the Medical University as well as the Institute of Molecular Pathology (IMP), which is Boehringer Ingelheim's cancer research centre. Furthermore, in 2004 the Institute of Molecular Biotechnology (IMBA) of the Austrian Academy of Sciences opened at the site focusing on functional genetics. Several young companies such as Intercell, MedSystems Diagnostics and VBC Genomics are also located here. Research focuses of the campus range from cancer therapy to the development of vaccines and state-of-the-art chip technology.

Another geographical and disciplinary nucleus has formed around the University of Natural Resources and Applied Life Sciences. The different departments are focusing on bioprocess research, applied genetics, microbiology and chemistry and today hold about 100 researchers, students and technicians. The campus houses companies like Polymun and Nano-S.

The General Hospital has since long been Vienna's central medical research establishment and employs more than 9 000 people. It serves as a home for 27 different university clinics and eight university institutes. It is a leading research hospital in areas like dermatology and cancer therapy and an important centre for clinical studies. Other hospitals in Vienna are also home to centres for clinical studies. Kaiser-Franz-Josef Hospital is a centre of excellence for clinical studies in oncology. Also, the Children's Cancer Research Institute is a multi-disciplinary competence centre dedicated to applying research and technology to improving recovery rates for cancers in children and juveniles. Recently, a new Research Centre for Molecular Medicine (CeMM) has been established. A new incubator building was finished in 2004 housing space for firms like Biomay and Fibrex Medical.

The Novartis Research Institute (NRI) was founded in 1970 and is situated in south of Vienna. Antibiotic Research Institute Vienna (ABRI) is another

 CLUSTERS, INNOVATION AND ENTREPRENEURSHIP – ISBN 978-92-64-04442-5 – © OECD 2009

Figure 4.3. **Research nuclei in the Vienna biotechnology cluster**

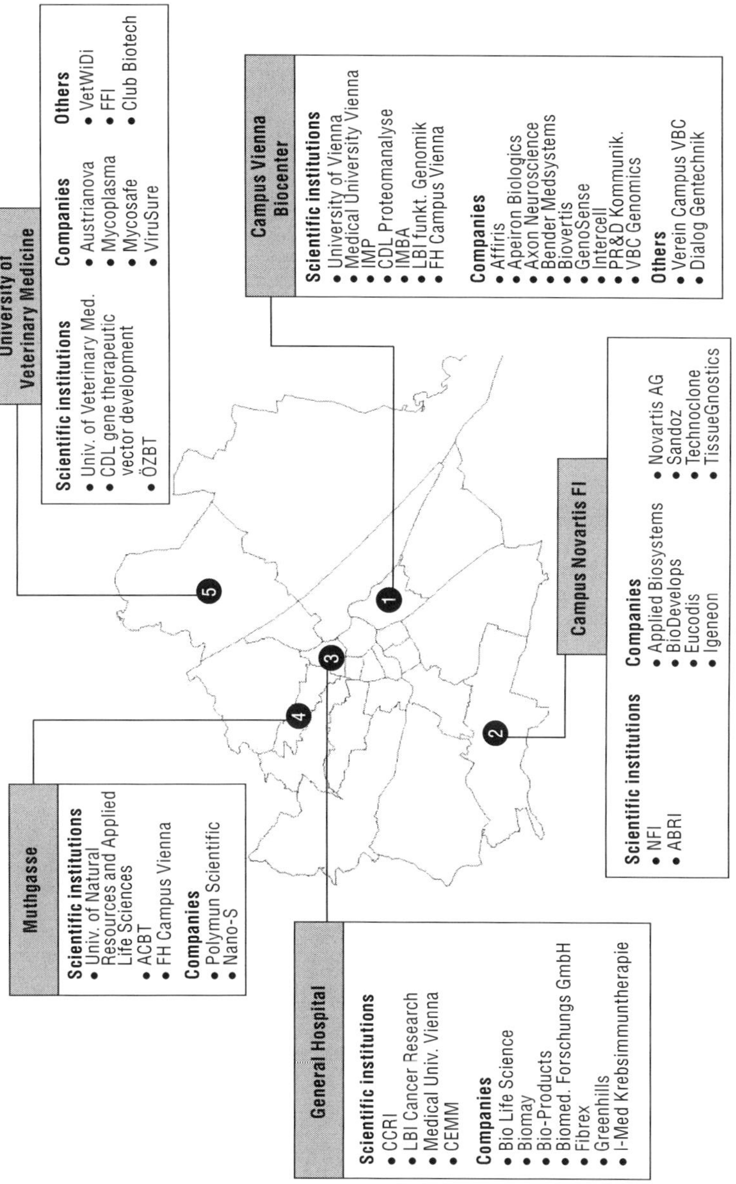

privately owned basic research institute owned by Biochemie Kundl (part of Sandoz R&D) located at the NRI since 2001. The NRI also houses space for start-ups like Igeneon and Technoclone.

The University of Veterinary Medicine forms another subcentre in the region. Research activities include the creation of transgenic animal models for use in the production of medicines and organs for transplantation. The firm Austrianova is also located at this site.

Besides research and provision of scientific knowledge, the universities located in the region of Vienna also fulfil an important function as a source of highly skilled labour. Moreover, the General Hospital Vienna is also home of the Vienna School of Clinical Research (VSCR) giving postgraduate training to physicians. In the recent past, the education and training system has become further differentiated. In order to meet the growing demand for skilled biotechnology technicians, several advanced technical colleges were founded. However, their alumni is not yet of considerable size.

Role of SMEs

Newly founded and small businesses play an outstanding role for the development of the biotechnology cluster in the region of Vienna. They could be regarded as crucial innovation actors, pointing to a general feature of the biotech industry per se, *i.e.* the prevalence of what has been termed an "entrepreneurial regime" in the literature (Audretsch, 1995; Fuchs and Krauss, 2003). The assertion of new small and medium-sized enterprises (SMEs) as important innovation agents holds true for spin-off firms in the Vienna biotechnology cluster in particular, which play a key role in transforming new knowledge into commercial products. An examination of the spin-off process in the Vienna biotechnology cluster reveals two main characteristics (see also Table 4.4):

- The overwhelming majority of all spin-out companies originated from academic institutions (university institutes, hospitals, other research organisations) operating in the region. Consequently, academic spin-offs are an important mechanism for the localised transfer of knowledge and expertise from the science system to the private sector. Moreover, most of them have established co-operative relationships with their incubators (see following subheading), thus, contributing to a strengthening and further enhancement of university-industry intersections in the cluster.
- The potential contribution of new firm spin-offs to the development of the Vienna biotechnology cluster, however, has not unfolded to a considerable extent so far. Looking at the age of companies and firm size we find that most spin-off companies are not older than five years and many of them are very small employing fewer than ten workers. Unsurprisingly, apart from a

 CLUSTERS, INNOVATION AND ENTREPRENEURSHIP – ISBN 978-92-64-04442-5 – © OECD 2009

few exceptions, they did not spin-off further second generation enterprises so far. Many spin-off firms are still in the phase of product development. Just a few companies have gone public so far and have reached a stage where they are earning revenues.

Table 4.4. **Characterisation of spin-offs in the sample**

		Number of firms	%
Age of firm	Not older than 5 years	9	60
	Not older than 10 years	4	27
	Older than 10 years	2	13
	Total	15	100
Location of parent organisation	Local	14	93
	National	0	0
	International	1	7
	Total	15	100
Type of parent organisation	Academic institution	11	73
	Firm	4	27
	Total	15	100
Firm size (number of employees)	1-10	8	53
	11-50	5	33
	More than 50	2	13
	Total	15	100

Source: Tödtling *et al.* (2006, p. 21).

Impact of the cluster on entrepreneurship and employment

The Vienna biotechnology cluster exerts a rather strong influence on the evolution of small enterprises and their innovation capabilities. In the following section we will concentrate on two specific issues that deserve further attention. These include the local labour market and the significance of the cluster as space for innovation related linkages. Finally, broader effects of the cluster for regional development will be discussed.

Labour market

The cluster's specialised labour market has been found to be crucial for the competitive edge of the large majority of the Vienna biotechnology firms. A closer look reveals that it is mainly the local universities with their continuous supply of highly skilled labour which play an outstanding role as provider of qualified personnel. The inter-firm mobility of talent at the local scale, however, is rather weak. Key for the further development of the cluster's labour market is the inflow of highly qualified scientists and managers from abroad. The influx of the latter group is of particular significance, as there is a lack of locally available managerial know-how, pointing to a central weakness of the cluster. Academic founders have only limited managerial competencies and top managers with high-level experience about the markets for pharmaceuticals and biotechnology are short in supply within the cluster.

A key space for knowledge interactions

The cluster represents an important space for knowledge interactions (see Table 4.5), enhancing the innovation capabilities of the firms present in the region (see also Tödtling and Trippl, 2007).

Table 4.5. **Types of knowledge links and their geography in the Vienna biotechnology cluster**

	Total	%	Networks			Spillovers and milieu			Market links			Other relations		
			With firms	With RO	Total	With firms	With RO	Total	With firms	With RO	Total	With firms	With RO	Total
Local	72	42	14	25	39	6	10	16	2	8	10	1	6	7
National	7	4	2	5	7	0	0	0	0	0	0	0	0	0
Global	93	54	17	17	34	15	9	24	13	7	20	7	8	15
	172	100			80			40			30			22

Note: RO ... research organisation (universities, clinics).
Source: based on Tödtling *et al.* (2006, p. 24).

More than 40 per cent of all knowledge links could be found within the Vienna region, pointing to a strong localised circulation of competences and expertise. An analysis of the nature of local knowledge links reveals particularly strong connections between research organisations and companies. These interactions are a multi-faceted phenomenon, reflecting the significance of different modes of knowledge transfer. Formal university-industry partnerships play a key role in this context. Some of them have been established with financial support of the national and regional governments, including the competence centre "BioMolecular Therapeutics" (BMT), the "Austrian Center for Biopharmaceutical Technology" (ACBT) and two Christian Doppler Laboratories (CD Labs). Furthermore, there is evidence of informal knowledge flows from academia to the local industry, underlining the importance of local spillovers and milieu effects.

Focusing specifically on the linkages between spin-off firms and their academic parent organisations uncovers rather vivid flows of knowledge, channelled by R&D co-operations, the joint use of infrastructure, the exchange of staff, the buying of patents and also the use of academic networks by the spin-off companies. Compared to contacts between academia and firms, local inter-firm relations are of minor significance in the Vienna biotechnology cluster. Some formal innovation networks between companies could be found and there is also some evidence of trust-based informal links between firms present in the region. It is, however, not only local knowledge interactions that matter. "Tapping-in" to global pools of expertise is highly relevant for the Vienna biotechnology cluster. More than 50 per cent of all knowledge interactions could be found at the international scale, pointing to the crucial significance of the inflow of knowledge from distant sources and partners. Formal networks represent a core mechanism for getting access to international sources of expertise. Additionally, the Vienna biotech firms benefit from global spillovers and milieu effects and have also established market linkages with international partners.

Broader effects for regional development

The Vienna biotechnology cluster is too young and too small in size to be an engine for regional employment growth. Nevertheless, it plays a key role for driving regional economic change. It could be argued that the emergence of the cluster has effects on regional development that go beyond the biotechnology sector. More specifically, the biotechnology cluster acts as a role model for the development of high-technology and knowledge-intensive industries in the region. The evolution of the Vienna biotechnology sector has been accompanied by a far reaching institutional change and the creation of a more favourable environment for seeding high-technology industries. New policy programmes, the creation of incubators, institutions fostering science-industry linkages, specialised structures for promoting high-technology firm formation, etc. all made their appearance in the past few years, leading to an "upgrading" of the regional innovation system. Breeding the biotechnology cluster a lot of experiences and expertise in nurturing high-technology clusters have been accumulated in academia, the business system and the policy community. Other knowledge-based sectors present in the region such as ICT or creative industries seem also to benefit from these qualitative improvements induced by the emergence of biotechnology in Vienna.

Barriers to cluster development

In the past years the Vienna biotechnology cluster has shown a positive development and entered into a catching-up process. In spite of that there are several barriers curtailing its dynamism and challenging its future evolution. The most important hindrances include the following factors:

- In international comparison the Vienna biotechnology cluster is rather small and in an early stage of its development. The cluster has not yet reached a sufficient critical mass and is still vulnerable.
- A crucial weakness of the cluster is the lack of home-grown Big Pharma companies which are regarded to constitute a key ingredient of successful biotechnology centres. There are also only few leading international companies present in the region, that go beyond simple sales and distribution activities.

- The cluster has failed so far to become a magnet for attracting other foreign companies. Since the 1980s no pharmaceutical companies carrying out research and production functions have located in the region. Furthermore, there is a rather low level of moving in of foreign biotechnology firms.
- A key barrier to the development of the Vienna biotechnology cluster is the bottleneck of venture capital at the local scale (see also Section 2). This does not only mean that financing opportunities are limited. It also points to a lack of functions which are usually fulfilled by venture capital firms in addition to the mere provision of capital. These include above all the selection of promising technologies and ventures, provision of support for the formation of management teams, the monitoring of the development of the young firms, establishment of contacts, etc.
- In the Vienna biotechnology cluster there is a lack of commercialisation expertise and managerial competences. Experienced managers with competences in the fields of pharmaceuticals and biotechnology are rare, and scientists are still not keen and used to commercialise their R&D results. Although a gradual rise of academic entrepreneurship is clearly observable in the cluster in the past years (see also Section 4), several factors continue to exert a hampering influence. Due to a weak entrepreneurial culture and a lack of experience, publication is still preferred to patent registration and new firm formation. These factors hamper scientists to take ideas or expertise they have developed in a knowledge institution and establish a new local business to further develop and exploit them.
- Finally, there is a low level of labour mobility among local firms. This leads us to note that there are still limits regarding localised knowledge circulation and collective learning processes, undermining the cluster's innovation capabilities.

Role of policy

There are three policy levels influencing the development of the biotech cluster in Vienna (see Figure 4.4). At the supranational level, EU programmes support and drive biotech related research through calls and funding. Moreover, legislation regarding biotechnology safe research as well as regulations of medical drug approval and clinical trials play governing roles. The Austrian policy and supporting system consists of national and regional institutions. Its strength lies in horizontal policies, which are not specifically designed to promote biotech but nevertheless have an impact on the development of its knowledge-base through calls, grants and funding of research institutions (Reiss *et al.*, 2003).

 CLUSTERS, INNOVATION AND ENTREPRENEURSHIP – ISBN 978-92-64-04442-5 – © OECD 2009

Figure 4.4. **Overview of the different policy systems governing Vienna biotechnology**

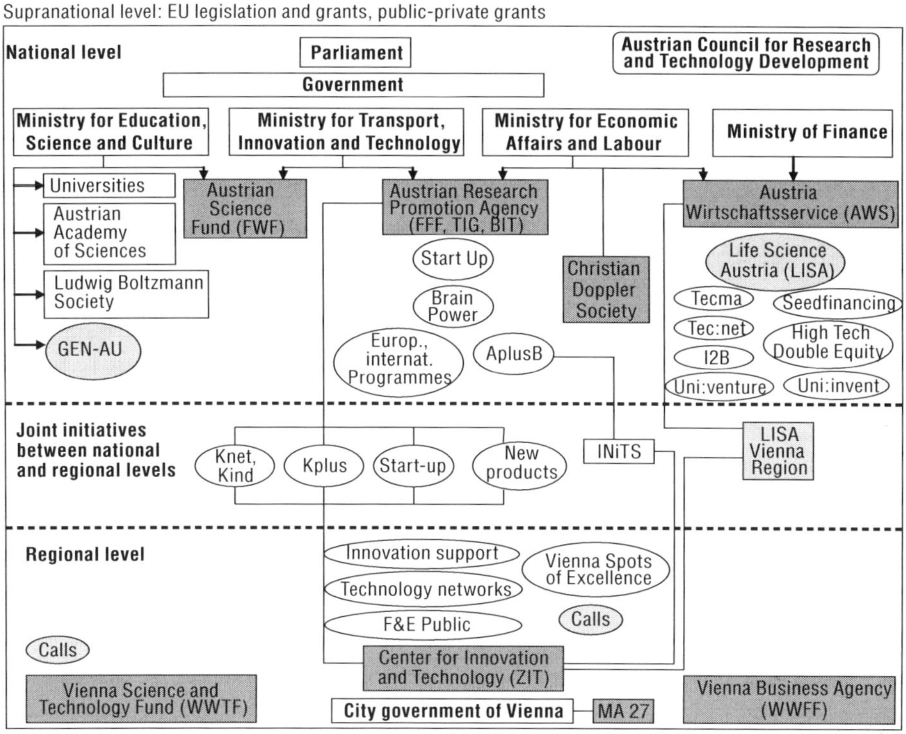

Source: Tödtling *et al.* (2006, p. 30).

National programmes

The national policy level plays an outstanding role in fostering the Vienna biotechnology cluster. There are a variety of different policy measures and instruments in this respect. Many of them, however, are not biotechnology specific but aim at encouraging technological development per se. There are just two programmes which specifically focus on the promotion of biotechnology. These include the Austrian Genome Research Programme (GEN-AU) and the initiative Life Science Austria (LISA). GEN-AU has been initiated in 2001 by the Ministry of Education, Science and Culture with a budget of EUR 10.5 million per year until 2010. It has the mission to strengthen genome research in Austria by funding interdisciplinary co-operative research projects undertaken by academic and/or industrial research teams. Consequently, its main focus is on facilitating dynamic and formal knowledge flows (networks) between regional and/or national partners. So far EUR 27.8 million have been allocated to 23 projects run by 91 partner organisations.

The majority of the funded institutions are located in Vienna, but a number of research teams in other Austrian provinces are receiving financial support, too. LISA is a programme of the Austrian Business Agency (AWS) launched in 1999 on behalf of two federal ministries. It mainly attempts to support the foundation of new biotech companies by providing financial support, advice and education. LISA consists of several components:

- LISA Pre-seed provides funds up to EUR 100 000 for a period of three years to potential entrepreneurs to establish the proof of principle.
- Moreover, LISA offers information and advice to firm founders with respect to technological and commercial issues.
- Best of Biotech (BOB) is a business plan competition promoting the entrepreneurial activity of researchers in bioscience related fields. Its goal is to increase the number of young life science firms by stimulating researchers to translate their ideas into business plans. BOB provides participants with advice and coaching with respect to prepare a business plan.
- Furthermore, lectures and training sessions are organised to enhance the commercial and managerial competencies of scientists.
- Finally, through the "Life Science Circles" meetings it also stimulates the informal exchange of ideas and experiences among actors. This is of particular importance in terms of enhancing the innovative milieu in the Vienna region.

Since 1999 about 35 new firm foundations in the field of life sciences have been supported, a large majority of which have settled down in the Vienna region.

At the national level we can find many other initiatives which were not specifically designed to promote biotechnology, they nevertheless contribute to its development. There are, for instance, several programmes set up by the AWS that aim to advance high-technology entrepreneurship. These include the initiatives "Seed financing" (loans), "High-Tech Double Equity" (acceptance of guarantees) and "uni:venture", which provides venture capital to academic spin-offs. Companies can receive up to EUR 1.1 million for a period of ten years. "uni:venture" contains a total funding volume of EUR 7.2 million.

Furthermore, there are the programmes "Tecma" and "uni:invent" by which the AWS supports universities, researchers and companies to patent and license their research results. This is done by providing expertise, training as well as funding support for patenting. Moreover it is engaged in searching for license deals.

Brain Power Austria is a programme carried out by the Austrian Research Promotion Agency (FFG) on behalf of the Ministry for Transport, Innovation and Technology. It has the goal to attract talented Austrian scientists from abroad. Scientists who are currently living or working in foreign countries are

CLUSTERS, INNOVATION AND ENTREPRENEURSHIP – ISBN 978-92-64-04442-5 – © OECD 2009

assisted in looking for career opportunities in Austria. The main activities in this respect include amongst others the provision of financial support, relocation services and coaching but also a promotion of Austrian job opportunities. Furthermore, the FFG actively promotes the participation of Austrian firms and research organisations in international co-operative RTD projects. More specifically it provides information and assistance relating to the Framework Programme of the EU, EUREKA and INTAS. Key activities include creating awareness, motivating, informing, and assisting on European Union and international research and technology activities, informing and coaching for preparing projects, as well as informing on issues of eligibility, evaluation criteria and procedural administrative and legal aspects. It also assists in searching for partners in collaborative EU and EUREKA projects. In addition, as the coordinator of the Innovation Relay Centre Austria, the FFG is actively involved in the transfer of new technologies and in other measures supporting innovation.

The Christian Doppler Society, founded in 1989, promotes the collaboration between universities, research institutions and industrial partners for a length up to seven years. Specifically, it has the aim of bridging basic and applied research in a certain area. Although it sets out for a more general programme, it is specifically targeting research in red biotechnology in the Vienna region.

Regional programmes

For a long time Vienna's economic policy was about providing subsidies to individual companies and attracting multinational firms. It was only by the end of the 1990s that a stronger focus on innovation and technology could be observed. Today, Vienna's strategic policy priorities are on life sciences, ICT, creative industries and the automotive sector. This reorientation of policy has been accompanied by a process of institution building. In 2000, the Centre for Innovation and Technology (ZIT) was established. Among its main activities are the funding of R&D activities of high-tech companies. One year later the Vienna Science and Technology Fund (WWTF) was set up to provide financial support to research organisations. Both funding agencies have special programmes for biotechnology organised as contests of proposals, thus following a "picking the winner" approach. Looking specifically at the promoting of knowledge links in biotechnology, the ZIT turns out to be the key institution. On the one hand, it provides infrastructure (laboratories, offices) to newly founded firms, thus advancing knowledge spillovers in form of spin-offs. On the other hand it has recently launched a comprehensive funding programme (ZIT 05 plus) that consists of several initiatives that are designed to be key drivers of knowledge links at the regional level:

- The initiative "Vienna Spots of Excellence" aims at promoting longer term university-industry partnerships.

- The new programme "Innovationssupport" provides funding for initiating and preparing partnerships with educational institutes as well as for co-operations in the fields of production, marketing and distribution.
- In order to support the formation and existence of networks in specific technological fields or centres, the programme "Technologienetzwerke" was launched. It provides funding for all networking activities that contribute to the success of the innovation network (*e.g.* information services, events, publications).

Joint programmes between regional and national policy levels

In the recent past, regional policy makers have managed to build up strong links with national actors, thus improving vertical coordination between the regional and national policy levels. The "Life Science Austria Vienna Region" (LISA VR) represents a good example in this respect, as it is a joint initiative between the ZIT and the AWS. LISA VR provides cluster management services to the local biotech industry and acts as a "one stop shop". By bundling the support available at federal and local levels, it offers a variety of services including consulting, pre-seed financing, education and mediation of incubation space. It also stimulates knowledge transfer as well as spin-offs. Furthermore it also promotes non-local formal and informal knowledge flows by participating in international fairs and promoting the cluster in relevant international media sources.

The national AplusB Programme has the task to support regions to establish centres that focus on the stimulation of new firm formation. In the case of Vienna this led to the creation of INITS. This centre has been founded in 2003 by the ZIT and two universities with the aim to promote technology-oriented spin-offs from the academic sector in Vienna by offering incubation space, counselling and assistance, specifically to academia, in the process of turning a good idea into a viable business.

The programmes Kplus and Knet/Kind focus on the formation of co-operative research centres between university institutes and companies. Both programmes were initiated in the second half of the 1990s by national ministries. Funding for the centres, however, comes not only from the national policy level but also from the regional one (in the case of Vienna, ZIT). The programmes demand a minimum of five partners and have a limit of seven years. Kplus has the general strategic goal of enhancing knowledge (basic research) within a specific discipline whereas Knet/Kind has the goal of innovating within the specific discipline (applied research).

Finally, the programme "Start Up", which is co-operatively run by the FFG and ZIT aims at supporting the formation of research intensive enterprises by funding R&D projects of young companies.

 CLUSTERS, INNOVATION AND ENTREPRENEURSHIP – ISBN 978-92-64-04442-5 – © OECD 2009

Table 4.6. **Regional and national policy programmes for promoting high-technology activities in general and biotechnology in particular (selection)**

Programme	Funding Agency	Aims: Promotion of ...	Instruments
NATIONAL PROGRAMMES:			
GEN-AU	BMBWK*	Co-operation	Financial support
LISA:	AWS		
Life Science Circles		Local interactions	Meetings, conferences
LISA Pre-seed		New firm formation	Financial support
Best of Biotech		New firm formation	Business plan competition
Tecma	AWS	Patents, licences	Advice, evaluation
uni:invent	AWS	Patents, licenses	Qualification of scouts
Seed financing	AWS	New firm formation	Loans
Uni-venture	AWS, BAWAG	New firm formation	Venture capital
i2b	FFG	Networks between business angels and firms	Brokering
Brain Power Austria	FFG	Inflow of highly qualified labour	Information, advice, financial support
Christian Doppler Labs	BMWA**	University-industry partnerships	Financial support
European and international programmes	FFG	R&D co-operations	Information, advice, brokering
REGIONAL PROGRAMMES:			
Infrastructure	ZIT	New firm formation	Provision of infrastructure
ZIT 05 plus:	ZIT		
Vienna Spot of Excellence		University-industry linkages	Financial support
Innovation Support		Co-operations	Financial support
Technology Networks		Networks	Financial support
Life Science Calls		R&D, innovation in single firms	Financial support
New Products		R&D, innovation in single firms	Financial support
F&E Public		Awareness raising	Financial support
Life Science Calls	WWTF	Scientific interactions	Financial support
JOINT PROGRAMMES BETWEEN REGIONAL AND NATIONAL LEVELS:			
LISA VR	AWS + ZIT	New firm formation, attraction of Foreign companies, etc.	Information, advice, location marketing
INITS (AplusB)	FFG + ZIT	New firm formation	Information, advice
Kplus, Knet/Kind	FFG + ZIT	University-industry linkages	Financial support
Start Up	FFG + ZIT	New firm formation	Financial support

* BMBWK Ministry of Education, Science and Culture.
** BMWA Ministry of Economic Affairs and Labour.
Source: Own compilation.

Overall, Vienna has been quite successful so far in achieving a certain cluster dynamics with relatively little money involved. The activities that have contributed to this success are:

- Vienna has used and combined quite effectively existing national (GENAU, LISA, BOB, AWS seed financing, "uni:venture") and new regional instruments (ZIT, WWTF Biotech Calls) for the development of its biotech cluster. It seems that there was a relatively good coordination between the national and regional level institutions.

- It has leveraged and stimulated private R&D investment through relatively small amounts of public money. Useful instruments have been horizontal programmes such as Kplus, Knet/Kind and CD laboratories.
- It has strengthened the cluster's scientific capabilities by setting up new specialised research organisations such as IMP and IMBA. Again, these were mainly nationally funded. IMP in particular was a public-private partnership between the Vienna University and Böhringer Ingelheim. A further positive effect has been the spatial concentration of science organisations and firms in specific centres of biomedical research such as the Vienna Biotech Center (VBC). This has allowed for a better knowledge exchange and knowledge spillovers.
- Vienna has used competitions ("calls" by the WWTF, ZIT, BOB) as a selection tool to support the most promising projects.

Vienna seems to have been less successful in the following fields so far, pointing to weaknesses and shortcomings in the overall set-up of policy actions and public efforts:

- The provision of finance for more capital intensive phases of product development, testing and firm growth remains limited. There is still a marked lack of policy measures dealing with these critical issues, which are central for the further development of the Vienna biotechnology cluster.
- Policy programmes geared towards attracting international risk finance and venture capital have not proved to be successful so far.
- The region of Vienna is not yet an internationally well known biotechnology location. This becomes clearly apparent if we consider the cluster's poor performance regarding the attraction of international big pharmaceutical companies and biotechnology firms. The outcome of policy measures which have been introduced to improve the location marketing is still modest.
- The Vienna biotechnology cluster is still very vulnerable and in a critical phase of development. The overall ensemble of policy actions has not been very effective to secure a stable growth of the cluster.

Policy adaptations over time

In the recent past biotechnology has attracted a lot of interest from policy makers in Austria and in the Vienna region. Compared to many other nations and regions, however, the support for this highly science-based sector comes late. As already outlined above, it was only by the end of the 1990s that systematic efforts to promote the biotechnology sector can be observed. This might be explained with the fact that Austria has no tradition and little

CLUSTERS, INNOVATION AND ENTREPRENEURSHIP – ISBN 978-92-64-04442-5 – © OECD 2009

experience in promoting high-tech industries. Whilst having a good research base, its national and regional innovation systems have not been ripe for breeding a strong and dynamic biotech sector as early and fast as it could be observed elsewhere. In the last years, however, an intensive stimulation of biotechnology has set in, brought about by a deep institutional change. There exists now a broad range of activities to foster start-ups and knowledge links in this sector, indicating that policy actors act like spiders in the web. This reorientation has been accompanied by a process of institution building: besides the Centre for Innovation and Technology (ZIT) and the Vienna Science and Technology Fund (WWTF) another new centre is INITS, which was founded in 2003. Its aim is to support technology-oriented spin-offs from the academic sector by offering counselling and assistance to scientists in the process of turning a good idea into a viable business. There are a variety of other organisations such as technology liaison offices at the universities. These supporting organisations are connected by informal relations. Regional policy makers have managed to build up strong links with national actors. Recently a joint initiative called "Life Science Austria Vienna Region" has started to provide cluster management services to the local biotech industry. Overall, the RIS has undergone a far reaching transformation. Key aspects in this context include the establishment of new research and educational organisations, which have strengthened the region's biotechnological capabilities, plus the creation of supporting agencies specialised in promoting high-technology industries and new policy routines.

- Looking at the types of instruments in use it is clearly visible that more traditional approaches such as funding and provision of infrastructure are still very important. They are, however, combined now with newer forms of intervention such as brokering, advice and cluster management services, resulting in a relatively balanced mix of older and newer modes of governance.
- It is obvious that national policy makers play a key role in the multi level governance system. Nevertheless, the regional policy level must not be neglected, as it plays a complementary role. Furthermore, there is a good vertical coordination in the policy and supporting system that manifests itself in the establishment of the initiative LISA Vienna region. Interestingly, a comparison of the policy actions undertaken at the regional and national level shows some unexpected results: national policy makers adopt a broader strategy, focusing on the promotion of many different types of knowledge (collaboration, informal contacts), new firm formation and international labour inflow etc. Furthermore, they fulfil various tasks such as financing and funding, brokering as well as provision of information and advice. At the regional policy level, in contrast, the focus is narrower, as new firm formation and formal networks are the key targets with funding as the

main instrument. This is somewhat astonishing as in most cases it is particularly the regional level where softer forms of intervention can be observed. We might conclude that policy makers in the Vienna region are addressing important barriers for the development of the biotechnology industry but are so far not using the full spectrum of available instruments.

Future policy challenges

Drawing on the insights presented in the last sections, we can identify the following future policy challenges for the Vienna biotechnology cluster (see Figure 4.5).

Figure 4.5. **Policy challenges for the further development of the Vienna biotechnology cluster**

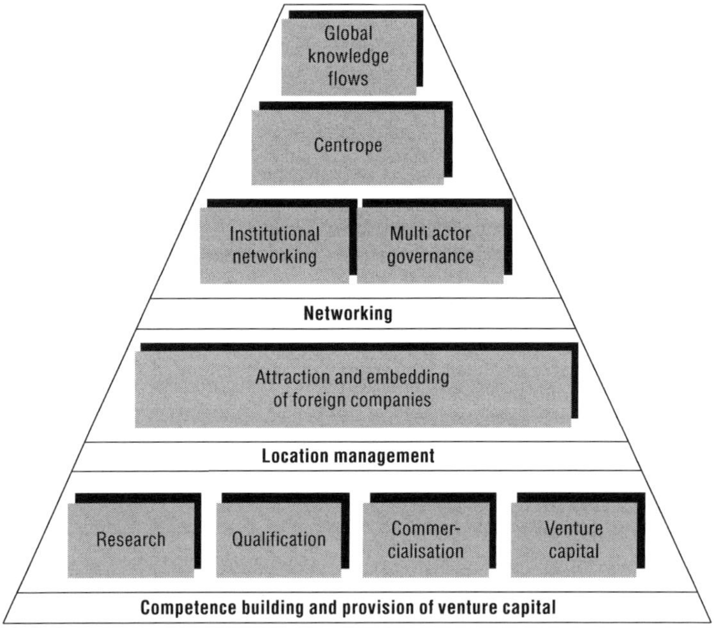

Source: Tödtling *et al.* (2006, p. 59).

Competence building and provision of venture capital

To stimulate the dynamic growth of the Vienna biotechnology cluster, continual investment in research, qualification and managerial competences as well as the provision of new forms of financing seem to be of vital importance:

- Further strengthening of research capabilities: excellent research institutes are regarded to be a key element of dynamic biotechnology clusters. In the past years, the knowledge infrastructure of the Vienna cluster has been substantially strengthened by the establishment of new research institutes. This strategy of accumulation of scientific excellence should be continued and intensified, as scientific knowledge is the central base for innovation in biotechnology.

- Qualification: The universities located in the Vienna region are a key source of highly qualified labour. In the recent past, the foundation of several technical colleges and the establishment of the Vienna school of clinical research have led to a further development of the educational system. The cluster's future evolution is critically dependent on further investment into human capital. The focus of activities should not be restricted to personnel with scientific capabilities but should also include labour with know-how regarding the commercialisation of scientific discoveries (see also below).
- Strengthening of managerial competencies: The Vienna biotechnology cluster suffers from a lack of experienced managers with competencies in the fields of pharmaceutical and biotechnology. To deal with this central problem, two policy approaches seem to be adequate. First, there is a need for measures to attract international top managers to the cluster. Second, to enhance the local commercialisation capabilities, special courses and seminars should be offered.
- Venture capital: Venture capital is the key missing ingredient in the cluster. To establish a public-private venture capital fund or to attract international venture capital firms should, therefore, be key objectives of the policy system.

Attraction and embedding of foreign companies

In the past years the Vienna biotechnology cluster has failed to attract foreign Big Pharma and young biotech firms, carrying out research and production in Vienna. This calls for new strategies in the field of location marketing and management, enabling to attract new companies to the region and to retain them in the cluster.

Networking

The relational dimension of the cluster also deserves further policy efforts. The most important measures in this context include:

- Intensification of institutional networking: The networking between regional and national policy organisations and supporting institutions (multi level governance) are well developed. In other areas of institutional networking, however, deficits prevail, demanding institutional innovations: First, the relationships between regional policy and supporting agencies are strongly informal in nature and do not go beyond an exchange of information and experiences. So far, no joint strategy for promoting the cluster has been developed. Second, the relations between the Austrian biotech centres Vienna, Styria and Tyrol are characterised by a strong competition. More co-operation between the provinces is called for to enable an exchange of information and experiences and joint policy learning processes. Third, a stronger networking between Vienna and other European biotech centres also seems to be a sound strategy.

- Multi actor governance: The process of strategy development for the cluster should also include representatives from the business sector and the science system. This accentuates the exigency of new institutional arrangements.
- Stimulating global knowledge interactions: The overall set of policy measures is characterised by a strong focus on promoting regional and national knowledge interactions. In comparison, only few policy initiatives are designed to explicitly stimulate international knowledge flows. To avoid the danger of lock-ins, more emphasis should be put on connecting the cluster with global knowledge sources.
- Intensification of networking within the Centrope region: The potential for a stronger networking between Vienna cluster actors and foreign companies and research facilities located in the Centrope region (formed by Vienna and neighbouring regions in the Czech Republic, Slovakia and Hungary) should be examined. If there is a critical mass of actors and sufficient synergies, measures geared towards the development of a cross-border biotechnology cluster should be adopted.

Lessons for other clusters

The emerging Vienna biotechnology cluster represents an interesting example for "seeding" high-technology industries in regional settings which have no tradition and little competencies for developing such sectors. The Vienna case challenges the prevailing view that the rise and evolution of such clusters are always spontaneous phenomena. The experiences of the Vienna biotechnology sector clearly demonstrate that there is a strong role of the state in facilitating the birth, early development and growth of high-technology clusters.

To promote biotechnology clusters in regions with weak potentials for high-technology industries, however, is a very complex and demanding process, requiring a large variety of policy measures and their sound combination. The most important activities in this context include to :

- Strengthen the cluster's scientific capabilities.
- Create new training and education centres.
- Ensure the availability of finance in general and risk capital in particular.
- Create organisations supporting entrepreneurship and the commercialisation of scientific knowledge such as academic spin-off centres.

 CLUSTERS, INNOVATION AND ENTREPRENEURSHIP – ISBN 978-92-64-04442-5 – © OECD 2009

- Attract innovative firms and talent.
- Promote knowledge interactions both within the cluster and beyond.
- Adapt framework conditions, etc.

Strengthening the clusters scientific capabilities requires the support of existing research organisations as well as the setting up of new ones. Public-private partnerships such as IMP seem to be a useful instrument to stimulate research activities with an application perspective. Collaborative R&D organisations such as Kplus centres and the support of collaborative R&D projects through programmes such as GENAU help to enhance the knowledge flows within the cluster and beyond.

New training and education centres are needed to provide the required qualifications and skills in a growing biotechnology cluster. Schools for both high level research and development staff (internationally well known universities) and for technicians and laboratory skills (technical colleges) are of vital importance.

Financial support should include different phases of projects and firm development: from R&D support and early phases of seed finance to the more mature and capital intensive phases of testing and product development. Both local and international venture capital firms are essential elements for such a financial support.

The Vienna and Austrian cases show quite clearly that programmes and policies can get more momentum and synergy if networks of public and private actors and supporting organisations can be formed. Such networks allow for the effective coordination of EU, national and regional programmes and instruments in order to stimulate cluster development.

Arguably, the development of a biotechnology cluster has to rely on a good endogenous R&D and innovation potential, *i.e.* local research organisations, firms, spin-offs and start-ups. But it cannot rely on local knowledge and entrepreneurial potential alone. The attraction of international innovative firms and of talent seems to be highly important for bringing in external competences and knowledge. Vienna's programmes and activities have not been so successful in this respect so far.

The spatial concentration of specialised research organisations and firms in specific biotechnology centres helps to stimulate local, often informal, knowledge flows and spillovers. Knowledge exchange should not be confined within the cluster and the region, however. In particular in the case of young and emerging clusters, such as in the Vienna case, the insertion into national and international networks and collaborations is of utmost importance. The support of collaborative projects at national and EU levels (framework programmes) is a useful tool in setting up such networks.

More often than not a renewal of the whole regional innovation system (RIS) and considerable institutional innovations are needed to effectively foster the emergence of biotechnology clusters in regional settings with weak conditions for high-technology sectors. RIS changes should become manifest in a broad set of areas, ranging from the field of hard infrastructure to socio-cultural factors such as new routines, conventions and attitudes. In terms of government modes it has to be emphasised that traditional instruments such as funding and the provision of infrastructure should be combined with modern forms of intervention, covering activities such as brokering, advice or cluster management services. Given the high level of complexity of such a transformation and the large number of policy agents involved, the success will critically depend on the interplay of different fields of policy at various spatial levels. There is a strong need of both horizontal and vertical policy coordination. The establishment of LISA Vienna region constitutes a good practical example for how to create an effective institutional arrangement for co-operation between regional and national policy makers. Vienna, however, has failed so far to provide the full range of elements necessary for a strong biotechnology cluster to emerge. Policies geared towards the development of a vivid venture capital scene have not been very effective in the past. Furthermore, a strong orientation of policy makers on promoting regional and national knowledge interactions could be observed and little has been done so far to encourage international knowledge flows which are of utmost importance for young biotechnology clusters in non high-technology regions. Finally, location policies have not been very successful to attract new innovative firms to the cluster. The key general lesson which could be drawn from the experiences of the Vienna biotechnology cluster is that considerable learning efforts and well co-ordinated actions in the policy community are necessary to facilitate the rise of high-technology clusters.

Notes

1. "Collective Learning in Knowledge Economies: Milieu or Market?" (2002-04) funded by the Austrian Science Fund; "Cluster development and policy in the Vienna biotechnology sector" (2005-06) funded by the Jubilee Fund of the City of Vienna for the Vienna University of Economics and Business Administration.
2. These numbers, however, refer to the larger Vienna region including the city of Vienna and the provinces Lower Austria and Burgenland and comprise not just "red" biotechnology but also other segments of the industry. Most of the employment, however, is concentrated in the Vienna region.
3. Starting with blood products, the company has profitably launched novel vaccines, went to the stock market in Zurich and then became acquired by Baxter in 1996.

CLUSTERS, INNOVATION AND ENTREPRENEURSHIP – ISBN 978-92-64-04442-5 – © OECD 2009

4. Funding for this project from the city of Vienna and the federal government was about EUR 10.6 million and can be seen as an initial step towards strengthening biotech research.

References

Asheim, B. and M. Gertler (2005), "The Geography of Innovation", in J. Fagerberg, D. Mowery and R. Nelson (eds.), *The Oxford Handbook of Innovation*, Oxford University Press, Oxford, pp. 291-317.

Audretsch, D. (1995), *Innovation and Industry Evolution*, MIT Press, Cambridge (MA).

Baier, B., E. Griessler and R. Martinsen (2000), *National Case Study of Austria. European Biotechnology Innovation System*, Institute for Advanced Studies, Vienna.

BIT (Bureau for International Research and Technology Co-operation) and LISA (Life Science Austria) (2004), *Bio-Tech in Austria*, BIT and LISA, Vienna.

Cooke, P. (2002), *Knowledge Economies. Clusters, Learning and Co-operative Advantage*, Routledge, London.

Feldman, M. (2000), "Where Science Comes to Life: University Bioscience, Commercial Spin-Offs, and Regional Economic Development", *Journal of Comparative Policy Analysis: Research and Practice*, Vol. 2, No. 3, pp. 345-361.

Fuchs, G. and G. Krauss (2003), "Biotechnology in Comparative Perspective", in G. Fuchs (ed.), *Biotechnology in Comparative Perspective*, Routledge, London, pp. 1-13.

Lawton-Smith, X. and X. Bagchi-Sen (2004), Guest editorial, "Innovation Geographies and Biotechnology", *Environment and Planning C*, Vol. 22, No. 2, pp. 159-160.

OECD (2004), *Biotechnology for Sustainable Growth and Development*, OECD, Paris, available at *www.oecd.org/dataoecd/43/2/33784888.pdf*.

Oliver, A. (2004),"Biotechnology Entrepreneurial Scientists and their Collaborations", *Research Policy*, Vol. 33, No. 4, pp. 583-597.

Oosterwijk, H., S. Rossak, and B. Unger (2003), "Austrian Biotechnology – Where to Find on the Map?" in F. van Waarden (ed.), *Bridging Ideas and Markets. National Systems of Innovation and the Organization of the Idea-Innovation Chain, Part II*, Country-Sector reports, final report of a project financed by the European Commission under the Fifth Framework Program (Targeted Socio-Economic Research), Utrecht University, Utrecht, pp. 203-231.

Reiss, T., *et al.* (2003), *Efficiency of Innovation Policies in High Technology Sectors in Europe (EPOHITE)*, final report, European Commission, Brussels.

Senker, J. (2004), "An Overview of Biotechnology Innovation in Europe: Firms, Demand, Government Policy and Research", in M. McKelvey, A. Rickne and J. Laage-Hellman (eds.), *The Economic Dynamics of Modern Biotechnology*, Edward Elgar, Cheltenham, pp. 99-132.

Technopolis (2006), Life Science – Standort im Vergleich, Studie im Auftrag der MA 27 EU-Strategie und Wirtschaftsentwicklung, Vienna.

Tödtling, F. and M. Trippl (2007), "Knowledge Links in High-Technology Industries: Markets, Networks or Milieu? The Case of the Vienna Biotechnology Cluster", *International Journal of Entrepreneurship and Innovation Management*, Vol. 7, No. 2, 3, and 4, pp. 345-365.

Tödtling, F., M. Trippl and J. von Gabain (2006), *Clusterentwicklung und -politik im Biotechnologiesektor Wien im Kontext internationaler Erfahrungen. Forschungsbericht*, Institut für Regional- und Umweltwirtschaft, Wirtschaftsuniversität Wien.

Trippl, M. and F. Tödtling (2007), "Developing Biotechnology Clusters in Non-High Technology Regions – The Case of Austria", *Industry and Innovation*, Vol. 14, No. 1, pp. 47-67.

Zucker, L., M. Darby and M. Brewer (1998), "Intellectual Human Capital and the Birth of the US Biotechnology Enterprises", *American Economic Review*, Vol. 88, No. 1, pp. 290-306.

 CLUSTERS, INNOVATION AND ENTREPRENEURSHIP – ISBN 978-92-64-04442-5 – © OECD 2009

ISBN 978-92-64-04442-5
Clusters, Innovation and Entrepreneurship

Chapter 5

The Life Science Cluster of Medicon Valley, Scandinavia

by
Professor Bjørn T. Asheim, Lars Coenen, Jerker Moodysson,
CIRCLE (Centre for Innovation Research and Competence
in the Learning Economy), Lund University, Sweden

This chapter provides a good example of a successful cross-regional cluster, and presents the challenges and benefits of such a structure. The engagement and the equal collaboration of the regions involved have been crucial in developing the cluster. Also, this chapter highlights the relevance of the "triple helix" model (industry, university, government) to stimulating innovation in a cluster, and the appropriateness of promoting a regional innovation system to dynamise the region. Moreover, this case shows and assesses the contribution of national and regional agencies for innovation in supporting the development of the cluster and encouraging collaboration.

Nature and evolution of the cluster

The life science cluster Medicon Valley are located in the bi-national Øresund region which spans greater Copenhagen in Denmark and Scania in southern Sweden, including the university town Lund and Sweden's third biggest city, Malmö (see Figure 5.1). In 2000, these two national parts were physically connected by the establishment of the 18 kilometre long Øresund fixed link (bridge and tunnel).

Figure 5.1. **Map of Medicon Valley**

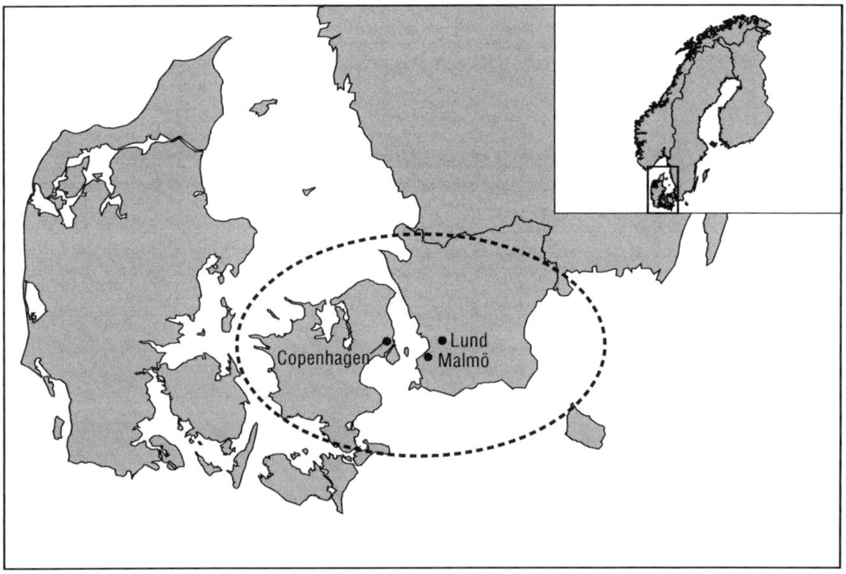

Source: Department of Social and Economic Geography at Lund University, Sweden.

In a recent study of the globalisation of biotechnology and life science industry, Phil Cooke (2005) identifies a hierarchy of globally networked bioregions in terms of size and level of innovation activities. A handful of US "megacenters" like Boston, New York and San Francisco are in the top of this chart, followed by European centres like Munich, Cambridge, Stockholm-Uppsala and Oxford. Medicon Valley can be considered a potential "megacenter" if seen as one bi-national cluster (see Table 5.1).

Table 5.1. **Comparative global performance indicators for bioregions**

Location	DBFs[1]	Life Scientists	VC (million USD)	Big Pharma Funding (million USD/annum)
Boston	141	4 980	601.5	800[2]
San Francisco	152	3 090	1 063.5	400[2]
New York	127	4 790	1 730.0	151.6[3]
Munich	120	8 000	400.0	54[4]
Medicon Valley	104	5 950	80.0	300[5]
San Diego	94	1 430	432.8	320[2]
Stockholm-Uppsala	87	2 998	90.0	250[5]
Washington DC	83	6 670	49.5	360[3]
Toronto	73	1 149	120.0	n.a.
Montreal	72	822	60.0	n.a.
Zurich	70	1 236	57.0	n.a.
Cambridge	54	2 650	250.0	105[3]
Oxford	46	3 250	120.0	70[3]

1. Dedicated biotechnology firms (DBFs).
2. Data 1996-2001.
3. Data 2000.
4. Data 2001.
5. Data 2002.

Source: Cooke, 2005.

Since the 1970s the life science sector has been led and dominated by large pharmaceutical companies (Big Pharma), producing and commercialising relatively few biotechnology based drugs. In the past decade the number of possible applications of biotechnology has multiplied, and Big Pharma is therefore increasingly dependent on new knowledge created by dedicated biotechnology firms (DBFs). In turn, DBFs are heavily dependent on the financial resources of Big Pharma (and venture capital) since the large up-front costs and long development times of new drug candidates require substantial cash-flow. Another important factor is the need for close relations with universities, research hospitals and other research organisations for intellectual property and knowledge inputs as well as the recruitment of skilled research personnel.

This "transformation" of biotech, with increased variety and complexity as major characteristics, has created new requirements for successful bioregions. The regions need to host a critical mass of strong actors representing not only the pharmaceutical industry but the whole value chain, provide suitable opportunities for basic research as well as commercialisation, and link up with knowledge sources in other bioregions across the globe. This report describes the development of the Medicon Valley biotech cluster over the last decade, both assessing how the cluster has adapted to these new requirements and identifying future challenges that the cluster faces.

Evolution of the cluster

The life science sector in Scania has long traditions through the presence of Astra (subsequently merged with Zeneca to become AstraZeneca) and Pharmacia (subsequently merged with Upjohn to become Pharmacia and Upjohn, and eventually acquired by Pfizer). Both these companies historically located significant research activities in Lund; AstraZeneca are still present with a major research unit employing 1 200 persons. After the Pharmacia merger the research on cancer and immunology was spun out to form the Lund based Active Biotech AB in 1997, while the rest of the company's activities disappeared from the region (the company still has a unit in Uppsala). Active Biotech AB is today, with 90 employees, the second largest and second oldest DBF in the region, after BioInvent International AB which today employs a staff of around 100 persons. BioInvent was created in 1995 by researchers at Lund University that wanted to commercialise their research. Besides these two medium-sized firms, the Swedish part of the region hosts about 35 other DBFs of varying size and age. A large share of the companies are university spin-offs (*e.g.* Camurus, Cellavision, Genovis and Wieslab) while others are local sub-units of global biotech companies (*e.g.* Acadia with headquarter in San Diego and research unit in Malmö).

Also the Danish part of the region has been a strong milieu for life science for a long time. Large anchor firms like Novo Nordisk and Lundbeck are still among the major players in the world, but local spin-off companies like Novozymes (research part of Novo Nordisk), local but world leading diagnostics companies like Dako (founded in Copenhagen 1966, today with sub-units in Colorado and California), and strongly associated pairs of complementary companies like Neurosearch (a biopharmaceuticals spin-off from Novo Nordisk) and NsGene (cell technology research spin-off from Neurosearch), have contributed to a renewal of the bioregion meeting new requirements on the global market. In total the Danish part of the region hosts about 100 DBFs.

The term Medicon Valley was first introduced in 1994 by the Øresund Comittee. This is a forum of public agencies from the Danish and Swedish part of the region with the mission to stimulate bi-national regional development. Feeding into the existing industrial specialisation of the region, it decided to focus specifically on the emerging field of biotechnology. Besides the historical localisation of Big Pharma (in fact, 60 per cent of Scandinavian pharmaceutical companies are located in Medicon Valley) an enormous potential for life sciences within the region was identified as it hosts 11 universities and 26 hospitals. However, the potential of becoming a global bioregion or "megacenter" are conditioned by the ability to achieve integration between the two national counterparts. This was hence the main ambition with the creation of Medicon Valley.

CLUSTERS, INNOVATION AND ENTREPRENEURSHIP – ISBN 978-92-64-04442-5 – © OECD 2009

The efforts to promote actual integration, making it justified to speak about Medicon Valley as one cross-border cluster (as opposed to two separate national clusters with less dignity on the global bio-market), took off for real with the formation of Medicon Valley Academy (MVA) in 1995 (in 1997, the organisation changed its name to Medicon Valley *Alliance*). MVA was initiated by Lund and Copenhagen Universities as an EU Interreg II project. The rationale behind the initiative was to stimulate the formation of a cross-border life science region by promoting local integration and cross-fertilisation between industry and academia. The MVA initiative has contributed to the development of the cluster, not the least because of its power of attraction on venture capital, research funds and human capital. This, together with the general transformation of biotechnology towards increased variety and complexity, has led to a shift in dominance from large pharmaceutical companies taking care of the entire value chain to small DBFs mainly focusing on basic research and early stages of development. Several of the large pharmaceutical companies have gradually downsized their production activities in the region, yet increasing their research facilities. At the same time there has been an impressive growth in number of DBFs. 65 new DBFs have been established since 1998, and if medical technology companies and R&D based service firms are included, the number of start-ups exceeds 100. Only in the period 2004-05, 29 new small R&D based firms were established in the region (MVA, 2006). Today there are approximately 130 DBFs of which approximately 70 per cent are located on the Danish side of Medicon Valley. Also university research, representing the earliest stages of the biotech value chain, has increased in the region.

This shift in dominance from single actors spanning the entire value chain to actors mainly representing the early stages has also affected the integration of the cluster and the needs for linking up with other bioregions. Actors in life science are today by necessity part of global research communities rather than regional ones. Due to their extreme specialisation they are forced to seek collaboration among the few potential partners available in the global arena, often only to be found in global biotech "megacenters" (Moodysson, 2007; Moodysson and Jonsson, 2007). For reasons like this, the initial enthusiasm over MVA as an initiative with the aim to strengthen local and cross-border integration has partly diminished. Several of the commercial actors gradually realised that "network promoting" activities without substantial output in terms of new formal collaboration were hard to justify, and academic actors felt a growing alienation against what they felt was more "the business of the local business" than something for them to engage in. As a result of this, MVA has adapted its strategy to meet the requirements of its members of a more dedicated focus on promoting global visibility of world class research. In its present "vision and mission" statement the focus has thus been broadened, not only to promote regional integration but also to "initiate synergetic collaboration with other bioregions and organisations and, together with others, promote and brand Medicon Valley, as well as the entire Øresund region, locally and globally" (MVA, 2006).

Recent examples of this strive to link up with other global biotech "megacenters" is the establishment of MVA "embassies" in places like Kobe, Vancouver, Seoul and Beijing, as well as the "UK-Medicon Valley Challenge Program" initiated in 2005. The aim is to develop world class biotechnology research and products by promoting research exchange and interaction between organisations in the Medicon Valley cluster and the biotech clusters in Cambridge, London, Liverpool-Manchester and Edinburgh. Examples of concrete activities within the programme are seminars, exchange of experiences between MVA board members and their UK counterparts, a joint EU 6th Framework Program including MVA and the Scottish Enterprise, and a UK-Medicon Valley Post-Doc Programme (MVA, 2006). The long-term vision of this collaboration is, according to the MVA chairman Per Belfrage to create "an air bridge from Medicon Valley to London and Cambridge, giving young scientists from Copenhagen and Lund the opportunity to experience these hot spots without having to move families and without having to worry about exorbitant housing prices".

Major actors of the cluster

Firms are key actors in the cluster as main drivers for innovation and industrial dynamics. According to MVA there are in total 130 biotech companies, 70 pharmaceutical companies and 130 medtech companies located in the region. Not all of these firms are engaged in or affected by research and development related to life science. When omitting those that only have sales or service departments in the region, or for other reasons cannot be classified as knowledge-intensive firms, the number is reduced to approximately 150 companies. Of these 150 firms, 130 can be classified as DBFs while the remaining 20 are either large pharmaceuticals or medtech firms. The ten largest firms in Medicon Valley are shown in Table 5.2.

Table 5.2. **Ten largest firms in Medicon Valley**

	Company	Empl.	Loc.
1	Novo Nordisk A/S	9 000	DK
2	H. Lundbeck A/S	2 100	DK
3	Coloplast A/S	1990	DK
4	Novozymes A/S	1 669	DK
5	Leo Pharmaceutical	1 270	DK
6	Unomedical A/S	1 200	DK
7	AstraZeneca R/D Lund	957	SE
8	Pfizer Health AB	850	SE
9	Radiometer A/S	847	DK
10	Chr. Hansen A/S	725	DK

CLUSTERS, INNOVATION AND ENTREPRENEURSHIP – ISBN 978-92-64-04442-5 – © OECD 2009

Universities are other important actors in Medicon Valley. Their role can be described according to three tasks. Firstly, they provide training and education to create and sustain a skilled pool of local researchers and scientists. Secondly, universities conduct publicly funded scientific research which can serve as knowledge input for DBFs. Finally, there is the so-called "third task" of universities which refers to direct collaboration between university and industry in the form of contract research as well as commercialising scientific research through licenses and start-ups of knowledge-intensive firms by university researchers. The most important universities in the region have been the universities of Lund and Copenhagen due to their long history of scientific excellence in medicine, biology and chemistry. Below we provide a general overview of the most important universities (and their departments) related to Medicon Valley.

Lund University was founded in 1666 and hosts eight faculties and a multitude of research centres and specialised institutes. It is today the largest unit for research and higher education in Sweden covering more or less all academic disciplines. The university has approximately 40 000 students and 6 000 employees. More than 3 000 post-graduates work at Lund University. Most doctorates are awarded in the medical sciences, followed closely by technology and the natural sciences. In 2006 the University had 581 professors. About half of all research at the University is externally financed. The most important research units for the Medicon Valley cluster are:

- *Faculty of Medicine* – In 2005 the faculty had 2 500 undergraduate students, 950 postgraduate students with 130 dissertations presented annually, while staff consisted of 139 professors and 450 other teachers and researchers. The faculty collaborates intensively with the university hospitals in Lund and Malmö to create an environment with productive communication between basic research and the healthcare system. One of the results of this collaboration is the Biomedical Centre (BMC).
- *Biomedical Centre* – This research centre, including the Stem Cell Centre and the Strategic Centre for Clinical Cancer Research (Create Health), is the largest single unit for teaching and research at Lund University, comprising half of the research at the Faculty of Medicine. It has a total of 700 scientists, including 50 affiliated professors, post-docs, Ph.D. students and technicians/administrative staffs working across 90 research groups. Major strongholds are today found in the fields of diabetes, immunology, neuroscience and cancer (BCG, 2002).
- *Faculty of Science* – In 2005 the faculty had 1 700 undergraduate and 440 postgraduate students with 85 doctoral dissertations presented. Staff consisted of 100 professors and 270 other teachers and researchers.

- *Lund Institute of Technology* – The Lund Institute of Technology, forming the engineering faculty of Lund University, was founded in 1961. It is Sweden's third largest institute of technology.

The formation of the Biomedical Centre (BMC) in 2001, initiated by Lund University to promote life science in the region, was underpinned by similar rationale as the MVA. The BMC assembles all the university's life science research under one roof, located adjacent to Lund University Hospital. This was primarily an attempt to rationalise the university research and strengthen the brand name of Lund University as centre of excellence in biomedical research. Hence, this initiative was mainly geared at strengthening the knowledge generation subsystem of the regional innovation system, while at the same time it contributed to promoting the integration of knowledge generation and early stages of knowledge exploitation. The concentration of related activities in one unit is completely in line with Lund University's building centres of excellence as part of a general development towards a more entrepreneurial university. The "flagship" of BMC is without doubt the Lund Strategic Research Center for Stem Cell Biology and Cell Therapy (Stem Cell Centre), established in 2003. Since the autumn 2006 BMC also houses a Bioincubator unit, which draws both on the concept of IDEON Incubation, and the services of Teknopol at (the immediately adjacent) IDEON, which was the first science park to be established in the Nordic countries in 1985, and on (the university-hospital hybrid) BMC as a source of new businesses, to extend the scope of commercialisation undertaken by the university to the active formation of biotech firms.

The University of Copenhagen was founded in 1479 and is the first university of Denmark. Spread over eight faculties from January 2007 after the integration of Danish University of Pharmaceutical Science and Royal Veterinary and Agricultural University as two new faculties, there are approximately 37 000 students and more than 7 000 employees. Except for management and engineering faculties, the University of Copenhagen qualifies as a broad, comprehensive university. Most relevant for Medicon Valley are the Faculties of Health Sciences (Medicine) and Science (as well as parts of the two new faculties). The University of Copenhagen has selected four Research Priority Areas for the years 2003 to 2007. The Research Priority Areas are set up to promote cross-faculty co-operation, encourage interdisciplinary research and education and strengthen the communication of research results and dialogue with society. One of these research areas is "Biocampus" targeting core biotechnological research.

- *Faculty of Health Sciences* – In 2005 the faculty had about 4 000 students of which 500 on the postgraduate level with 160 dissertations presented. Staff consisted of approximately 60 professors and 400 other teachers and researchers. Active collaboration takes place between the faculty and BRIC – the Biotech Research and Innovation Centre and the University Hospitals.

CLUSTERS, INNOVATION AND ENTREPRENEURSHIP – ISBN 978-92-64-04442-5 – © OECD 2009

- *BRIC* – The Danish Ministry of Science, Technology and Innovation established BRIC to form an international centre of excellence in molecular biology. BRIC is a consortium of leading Danish research institutions and is associated with the University of Copenhagen. One of its missions is to establish collaborations between public research institutions and industry, and to promote exchange of ideas within the Danish biotech research community. To achieve this, BRIC is managed by an independent board of directors representing academic institutions as well as industry. BRIC has now 5 research groups and a total of 40 employees.
- *Faculty of Science* – In 2005 the faculty had approximately 6 000 students of which 317 on the post-graduate level with 85 dissertations presented. Staff consisted of approximately 90 professors and 560 other teachers and researchers.

In addition to firms and universities, research institutes play an important role for basic research and discovery. The most important institutes in Medicon Valley are:

Carlsberg Research Center is an independent private research centre and part of Carlsberg A/S. Traditional competences are in malting, brewing and fermentation but it has become increasingly active in biotechnological production processes and biomedical sciences to target early drug discovery.

The Hagedorn Research Institute is an independent basic research component within Novo Nordisk A/S in the field of diabetes and its complications. The three main areas of research activity are 1) stem cell research and developmental biology of the pancreas 2) immunology and genetics of Type 1 Diabetes and 3) structural and systems biology of ligands and receptors of the insulin peptide family. The institute also fulfils an educational mission by training a substantial number of masters and PhD students in collaboration with Danish universities, the Ministry of Science, Technology and Innovation (VTU) and the Medicon Valley Academy (MVA).

AstraZeneca's respiratory research unit in Lund, with approximately 800 research employees, accounts for roughly two-thirds of inflammation and immunology R&D in Medicon Valley.

IHE, the Swedish Institute for Health Economics is a non-profit research institute that has been operating in Lund since 1979 and is a wholly-owned subsidiary of Apoteket AB (the National Corporation of Swedish Pharmacies).

SIK – the Swedish Institute for Food and Biotechnology has a regional office in Lund. SIK conducts strategic and applied research within food and biotechnology in accordance with an industry-governed, goal-oriented research programme and through joint industrial projects.

Statens Serum Institut is a public enterprise operating as a market-oriented production and service enterprise under the Danish Ministry of the Interior and Health. The institute aims to ensure advanced control of infectious diseases, including new infections and biological threats and conducts state-of-the-art research in this field.

The *Danish Cancer Society* has two dedicated institutes both located in Copenhagen: the Institute of Cancer Biology and the Institute of Cancer Epidemiology.

Health care institutions supply expert assistance and research in different aspects of drug and device development. The most important hospitals are Copenhagen Hospital Corporation, Copenhagen County Hospital, Lund University Hospital and Malmö University Hospital.

Success factors

Venture capital

Financial backing is essential for the development and growth of a biotech cluster. Venture capital is needed to facilitate entrepreneurship and commercialisation of scientific research. MVA identifies 41 relevant investors for venture capital with offices inside (61 per cent) as well as outside the region (39 per cent). This is due to the fact that the Danish investors are primarily located in Copenhagen while the Swedish investors are predominantly concentrated in Stockholm. 50 per cent of all investors can be found in Copenhagen against only 10 per cent in Malmö or Lund.

Figure 5.2 provides an overview of the distribution of financial specialisation in terms of investment stages. It shows that most emphasis lies on seed and start-up capital while pre-seed and restructuring receive relatively less attention. However, the tendency for 2005 was that more investments were made in mature companies and less in start-ups, which makes somewhat of a trend shift compared to the figures presented below (MVA, 2006b).

The national venture funds are especially important players for Medicon Valley even though they can only invest in their own country. In Denmark these are Vaekstfonden and Dansk Innovationsinvestering, both located in Copenhagen. During 2003 Vaekstfonden made investments in biotech companies for EUR 10 million and Dansk Innovationsinvestring for EUR 7 million. The Swedish Industrifonden, which is located in Stockholm but with a regional office in Malmö, has invested EUR 4 million in Medicon Valley. Another important category is the venture capitalists affiliated with the incubator centres in Medicon Valley: DTU Innovation (connected to the Danish Technical University), CAT-Symbion Innovation and Teknologisk Innovation in Denmark and TeknoSeed in Lund. Finally, private seed investors include Novo, the biotech fund for Novo Nordisk and LEO incubator.

CLUSTERS, INNOVATION AND ENTREPRENEURSHIP – ISBN 978-92-64-04442-5 – © OECD 2009

Figure 5.2. **Distribution venture capital Medicon Valley**

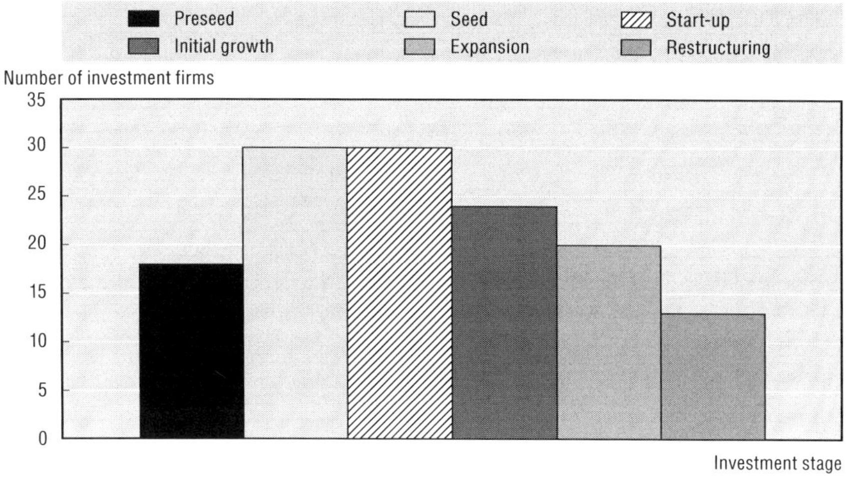

Note: One investment firm can cover multiple investment stages.
Source: MVA (2001).

Network organisations

Network organisations have proven to be key venues and meeting grounds that provide the social platforms to exploit the opportunities of co-location in a cluster.

Medicon Valley Alliance (MVA): The largest and probably most important network organisation for Medicon Valley with 280 members (counted January 2008). If any, MVA should be considered as *the* cluster organisation. As a member financed network organisation it works to promote the necessary interaction for network formation and knowledge transfer between academia, public health, and biotech related industries. Important tools in this are seminars and conferences, as well as initiating and coordinating projects associated with educational, scientific and business activities in the region. MVA also sets up and manages comprehensive knowledge databases and has initiated a range of working groups to analyze regional competences within specific subject areas. In addition, MVA contributes to the regional and international marketing of Medicon Valley by visiting and presenting the cluster at conferences and other events and, as mention above, by establishing "embassies" in other important bioregions globally.

MVA is a non-profit association predominantly based on revenues generated through membership fees. In 2005, these constituted 75 per cent (Denmark, 52 per cent and Sweden 23 per cent) of the total annual turnover of approximately EUR 1.2 million. Forty-seven per cent of the total membership fee is paid by private funds while 53 per cent is funded publicly. The remaining

25 per cent were accounted for by a contribution from the Øresund Science Region (13 per cent), PhD administration (5 per cent), sponsorships and funds (4 per cent), a VINNOVA project (2 per cent), and seminar fees (1 per cent). In total, MVA has a staff of 10 people responsible for daily operations and a board of directors of 15 representing the different members of the organisation. The board of directors is elected at the annual general assembly.

Øresund University: This is a consortium of fourteen universities and university colleges in the Øresund region with the objective to increases quality and efficiency among the participating institutions by opening up all courses, libraries and other facilities to all students, teachers and researchers. Of the fourteen universities, four are Swedish and ten are Danish. The number of Danish and Swedish students, however, is about the same. Øresund University is, similar to MVA, part of Øresund Science Region.

Øresund Science Region: ØSR joins the forces of six regional research and innovation platforms, Øresund University and a number of regional co-ordination bodies in an attempt to strengthen regional co-operation and integration between universities, industry and the public sector. The six ØSR platforms are: Medicon Valley Academy; Øresund IT Academy; Øresund Food Network; Øresund Environment Academy; Øresund Logistics; Øresund Design. The activities of the platforms include establishing partnerships, benchmarking, enhancing research and education, innovation, technology transfer and marketing.

Role of SMEs

Enterprises

In a number of enterprises, the Danish side of the region has historically dominated. Eighty-six per cent of the region's medico/health enterprises were located here in 1995. However, since the mid 1990's the number of enterprises has increased significantly in Sweden, while Denmark has experienced a slight decrease. In 2004 Øresund hosted about 4 300 private enterprises within the field of medico/health. 255 of those are in the fields of biotechnology and pharmaceuticals, and 235 are specialised in research and development. Seventy-five per cent of all enterprises are located on the Swedish side (Figure 5.3), but within the subgroup of biotechnology and pharmaceuticals, 59 per cent are still located in the Danish part (Vinnova, 2005b).

Analyzing this change in relation to the employment figures and developments on the national levels, it becomes evident that this does not reflect a major catch-up of Scania. What happened on an aggregated level is that while the Swedish firms have become more in numbers but smaller in size, the Danish have become fewer and larger. Similar changes seem to have taken place on a national level. Still, similar to 1995, Øresund hosts about

CLUSTERS, INNOVATION AND ENTREPRENEURSHIP – ISBN 978-92-64-04442-5 – © OECD 2009

Figure 5.3. **Number of medico/health enterprises (private sector)**

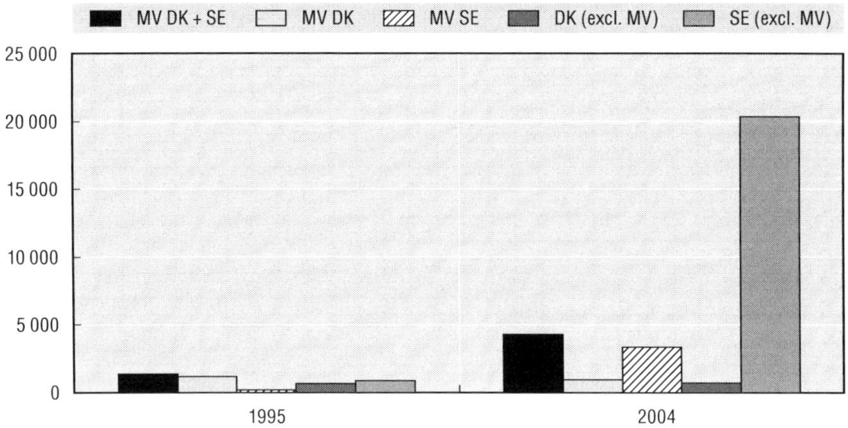

Source: Ørestat (2007).

15 per cent of all Sweden's medico/health firms, compared to 60 per cent of all Denmark's. We can trace no change over time in these relative figures. In the biotechnology sub-sector about 25 per cent of all Swedish firms are located in Øresund (Vinnova, 2001). In Denmark, 75 per cent of all biotech firms are located in and around Copenhagen (Bloch, 2004).

Looking at the size structure of the life science enterprises in the region (both public and private sector, but excluding hospitals) we can see that small units dominate (Figure 5.4). About 90 per cent of all local units have less than 50 employees. This is not surprising however, since the calculations are based on numbers of local units, not on their share of total employment. What is

Figure 5.4. **Size structure of local medico/health units in Øresund**

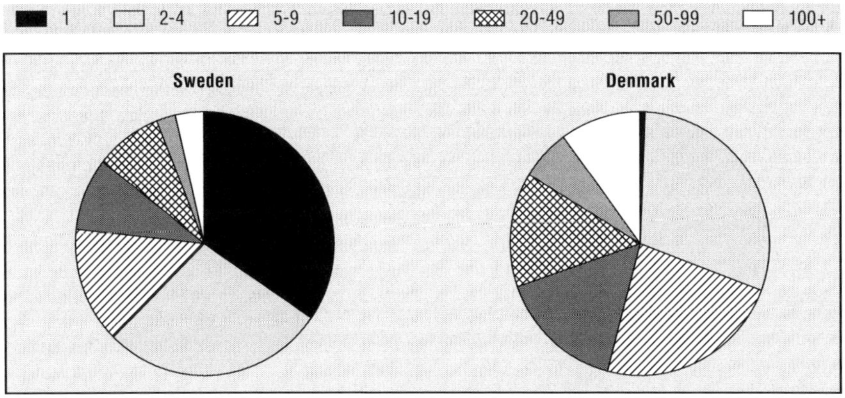

Source: Ørestat, 2007.

striking though, and in accordance with the observed changes in number of firms, is that the size structure differs significantly between the Swedish and Danish part of Øresund. In Sweden, we find a large share of single-employee firms (publicly employed researchers or doctors with their own company on the side), while these are hardly represented in the Danish part. The three larger groups in the classification (local units with 20-49, 50-99 and more than 100 employees) are on the other hand far more represented in the Danish part of Øresund (Figure 5.4). When it comes to the really large pharmaceutical firms, only AstraZeneca are located in the Swedish part, compared to more than ten companies in Denmark (Vinnova, 2005a).

Impact of the cluster on entrepreneurship and employment

Economic performance

The total life science turnover in the region has increased from EUR 6.9 to EUR 14.8 billion between 1995 and 2004 (Figure 5.5), while value-added has increased from EUR 3.3 to about EUR 4.9 billion in the same period (Figure 5.6). This may appear as strong growth, but corresponding figures for Sweden and Denmark as a whole, EUR 16.4-39.3 respective EUR 9.3-14 billion, put these figures in perspective. As in the case with employment, the Danish part of Øresund represents a large share of Denmark's total turnover (83 per cent) and value-added (88 per cent), while the Swedish part is far below the country as a whole. In 2004, the Scanian share of Swedish medico/health turnover and value-added was about 8 per cent. However, it is not primarily in purely economic terms that Medicon Valley stands out as a success story.

Figure 5.5. **Turnover medico/health**

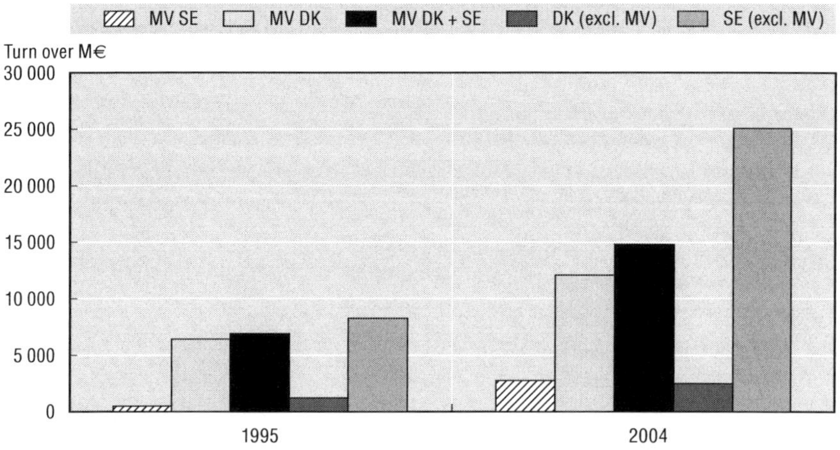

Source: Ørestat, 2007.

Figure 5.6. **Value-added medico/health**

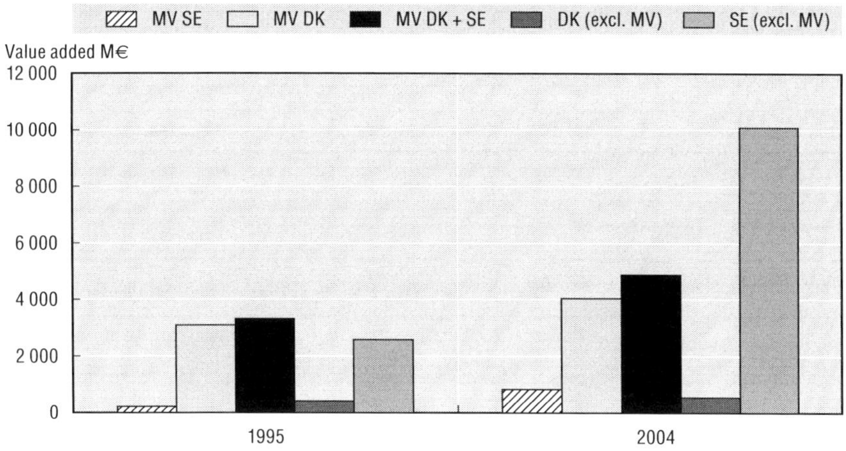

Source: Ørestat, 2007.

Employment

In a region with about 3.5 million inhabitants, the resource area medico/ health employs 50 000 people (full time equivalent numbers). This means a growth of nearly 15 000 employees since the mid 1990's (Figure 5.7). Thirty-five thousand are employed in companies active in research and development in the field of biotechnology, medtech and pharmaceuticals (Vinnova, 2005a).

Figure 5.7. **Medico/health employment (private sector)**

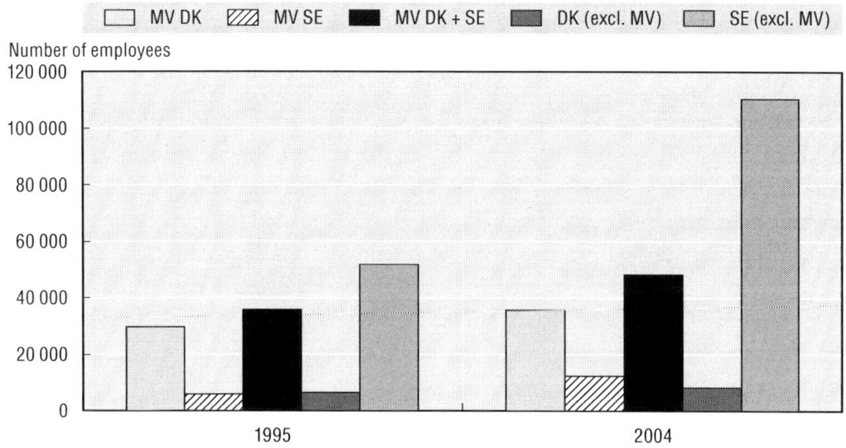

Source: Ørestat, 2007.

More than 5 000 private sector employees are trained researchers (*i.e.* hold a PhD degree). If public sector academic research is included, the number of researchers adds up to more than 10 000. In addition, the region hosts about 2 000 PhD students within the field of life science, affiliated to the region's universities and university hospitals (MVA, 2004).

The employment distribution in Øresund is quite unequal from a national perspective. Eighty-three per cent of all medico/health (as measured in full time employment) takes place on the Danish side. Also in relation to overall national figures, the Danish part stands out as particularly strong. In 2004 more than 80 per cent of all Denmark's medico/health activity was concentrated in the Greater Copenhagen region. In Sweden, Scania accounts for about 13 per cent of all Sweden's medico/health, which makes it the third most important life science node in the Swedish economy after Stockholm-Uppsala and Gothenburg (Vinnova, 2005b). This skewed distribution can partly be explained by a Swedish specialisation in the small-firm dominated and labour extensive sub-sector of "dedicated biotechnology" whereas the Danish side has more large pharmaceutical companies.

Entrepreneurship

Even though the local presence of Big Pharma in Medicon Valley should not be underestimated as a key and, to a certain extent, unique cluster asset, it is equally important that an innovative milieu is maintained for the start-up and growth of smaller DBFs. Globalisation processes in life sciences remind us of the volatile business behavior of large pharmaceuticals (*e.g.* Novartis decision to relocate its headquarters from Basel to Boston). Current presence of Big Pharma is by no means guaranteed on the long run. Nonetheless, the local node global network pattern implies that there are certain factors that favor concentration of biotechnology. These factors are more strongly connected to local academia and the presence of DBFs than to Big Pharma. It seems however that the region does not entirely reap the commercial potential of this. The strategic report by Medicon Valley Academy (2004) on this issue, called "From Bioscience to New Jobs" confirms the region's strength in scientific research but remains critical about the commercialisation performance of Medicon Valley universities compared to other prominent bioregions such as Stockholm-Uppsala, Oxford and Stanford. While the number of invention disclosures leading to a patent is high for academic researchers in Medicon Valley, the rate by which these disclosures result in license agreements or firm start-ups is disappointingly low. On a general level, the number of start-up companies is however high. This outcome can be explained by a specific weakness in the academic system to commercialise research.

In Medicon Valley the majority of the science parks and their incubator facilities are independent, market-driven organisations and, as such, only

 CLUSTERS, INNOVATION AND ENTREPRENEURSHIP – ISBN 978-92-64-04442-5 – © OECD 2009

loosely connected to the universities and their technology transfer offices. Only IDEON in Lund and SCION-DTU in Copenhagen serve as an exception. This may hamper a close relationship between academic researchers and the various incubators. This in turn can partly explain why start-ups in Medicon Valley seem to originate more often from employees having previously worked in a corporate setting than from university researchers.

There is also a major difference between the ways commercialisation of academic research is regulated in the two countries. In 2000, Denmark adopted a law similar to the well-known Bayh-Dole act in the US which allows public research institutions (universities and hospitals) to claim the commercial rights of inventions from their employees. Disclosure is therefore a legal requirement in the Danish system. Following the adoption of this law, each university and university hospital established a local technology transfer office. Commercial revenues of an invention (in the form of license agreement with an existing company or by transferring these rights to a new firm start-up) are divided along the rule: one third goes to the inventor, one third goes to the tech transfer office, and one third to the university.

According to the Swedish "teachers exception", university researchers are allowed to keep the intellectual property rights of invention and, in contrast to Denmark, there is no obligation to disclose new findings to the university. There is thus more individual freedom for an inventor to choose which route to use for commercialisation. Technology transfer offices at university are therefore less widespread than in Denmark. However, somewhat similar organisations to the university's own technology transfer offices exist in the form of the "Technology Bridge Foundation" organisation. This is a foundation which has been set up by Swedish parliament and which is organised and co-ordinated on a regional basis. The office for Scania (which is also covering four other regions in southern Sweden) is located in Lund. The Technology Bridge Foundation in Lund has separate divisions for evaluation and counseling (Teknopol), patents and licenses (Forskarpatent i Syd) and seed investment (TeknoSeed). A difference between the university based technology transfer offices and the science-park based Technology Bridge Foundation is that the former are considered to be more proactive in their efforts to scout for and commercialise new research findings. Still, according to the MVA report (2004) these technology transfer offices (in both Sweden and Denmark) have strong competencies in mainly science, law and regulations but relatively little experience from industry (as compared to technology transfer offices in the US or UK). It is indicated that a lot of the resources have been used for patents that are not very commercially attractive or were not properly marketed. These offices are fairly small with limited staff compared to similar offices in other global bioregions. In San Francisco, Boston and Oxford, the number of employees is between 24 and 35 while the average size in Medicon Valley is 3. This is however not surprising given the novelty of this institution in Scandinavia. Finally, several of the large Danish pharmaceuticals have technology incubator facilities in place but these are mainly directed at spin-outs from their own R&D facilities.

Due to the local presence of many private venture capital firms, national public seed investors (particularly in Copenhagen) and the various incubator facilities, the general venture capital climate in Medicon Valley is considered to be good. However, due to the fact that Copenhagen, the financial heart of Denmark is located on the Danish side of Medicon Valley, the distribution of investment capital is skewed. Moreover, the division of the region in two national jurisdictions further aggravates this situation. National public seed investors are allowed to only invest in national ventures. Private venture companies are less restricted in this respect. However, in order to stimulate commercialisation of university research more pre-seed venture capital would be needed. Secondly, initial growth conditions for novel biotech firms are considered to be less beneficial for Medicon Valley in terms of funding support.

Barriers to cluster development

The cluster is unique in being located in a region that spans parts of two different countries. This feature can and should not be ignored as a weakness of the cluster compared to similar life science clusters in scope and size such as nearby Stockholm-Uppsala. The national border contains a liability for fragmentation of cluster activities. This report shows that the full potential for synergy effects derived from close proximity may be hampered because of this. In other words, cross-border regional interaction is only weakly developed. In addition, extensive parts of the regulative environment (*e.g.* tax-rules, employment legislation) as well as research and innovation policy is shaped and implemented within a national framework which complicates cross-border cluster interaction.

Part of the explanation of cluster fragmentation, can be the imbalance between the different parts of Medicon Valley. The Danish side is strongly characterised by being the capital city region and the only major urban area in the country. More or less by default, there is a lot of business activity going on in terms of financial markets, presence of business headquarters as well as political activity. Scania, on the Swedish side, is typically a second tier region being relatively far away from the action in the capital Stockholm. In terms of inhabitants, number of firms (especially Big Pharma), and venture capital investors, there is clearly a skewed distribution in favor of the Danish part of Medicon Valley. As for university and research facilities however, the situation is more balanced.

Another potential barrier to the development of the cluster is the somewhat disappointing results in commercialising academic research especially compared to North-American life science clusters. Measured in terms of establishing start-

 CLUSTERS, INNOVATION AND ENTREPRENEURSHIP – ISBN 978-92-64-04442-5 – © OECD 2009

up companies or license agreements following patent filings, the Medicon Valley universities (and university hospitals) fall behind important competitors such as Karolinska Institute in Stockholm, Oxford University and Stanford University alone. Partly, this can be explained by the fact that "the entrepreneurial university" has its origin in the US while Europe in general (with the exception of the UK) is following suit. Therefore, it should be emphasised that legislation and the establishment of technology transfer offices is still in the process of catching up and considerable progress is being made. A lot of policy efforts can be expected to contribute to this. Another disadvantage, at least compared to the North American life science clusters, is the lower supply of venture capital.

Role of policy

Both countries have thriving innovation policy environments. The main governmental innovation body in Sweden is VINNOVA (Swedish Agency for Innovation Systems) which was established in 2001. It has an annual budget of about EUR 110 million to support innovation on a national, regional and sectoral level in active collaboration with industry and academia (following the "triple helix" model). Life sciences are broadly covered in four so-called growth areas: "drugs and diagnostics", "biotechnical tools", "medical technology" and "innovative food". It can therefore be seen as a prioritised technological platform in Swedish innovation policy not the least because it receives approximately one tenth of the total annual budget. Moreover, VINNOVA has committed itself to support the absorption of biotechnology in the food sector in Scania through its regional innovation systems programme VINNVÄXT. In Denmark, innovation policy is co-ordinated through the Ministry of Science, Technology and Innovation (VTU). Compared to Sweden, there is less explicit state support and involvement for innovation or "triple helix" collaboration. VTU supports however "innovation consortia" to enhance co-operation between public institutions and private enterprises. An example of this is the recently established Danish Pharma Consortium under initiative of four Danish Medicon Valley universities (Copenhagen University, Danish Technical University, Danish University of Pharmeceutical Science and the Royal Veterinary and Agricultural University). Another important component of Danish innovation policy are the 15 national Business Service Centres (*Erhvervsservicecentre*) which provide counselling and information to SMEs at a local level. The aforementioned incubator facilities should also be mentioned as part of both countries' innovation policy. In sum, therefore, it needs to be emphasised that the innovation policy environment for Medicon Valley is very strongly divided along national boundaries. There is no systematic collaboration or coordination between VINNOVA and VTU for Medicon Valley.

Even if Medicon Valley is not a direct result of national or regional political initiatives, it has indirectly benefited from the favourable political environment for supporting high-tech as well as cluster development. Sweden has for many years pursued an active innovation policy through the national responsible agency VINNOVA. This has been a combination of technology policies, supporting specific, strategic technologies and sectors – of which biotech was/is considered to be one of the most important, and innovation policies, promoting the formation of regional innovation systems, primarily for supporting high-tech industries. So far, out of eight Vinnväxt projects, three belong to the biotech sector (two red bio in Gothenburg and Uppsala, and one green bio in Lund).

Denmark on the other side has a tradition of less direct public intervention, and leaves more to the market. Thus, Denmark, which in contrast to Sweden, is dominated by SMEs, is characterised by a market-driven innovation system supporting non-R&D based, incremental consumer product innovations. One of the exceptions to this is the pharmaceutical industry, which is research intensive and dominated by large companies. The support of this industry mainly takes the form of science policies of funding basic research at universities and research institutes, even if it amounts to less than the similar funding in Sweden.

As has already been stated in the report, the establishment of Medicon Valley Academy/Alliance is a result of an Interreg project initiated by Lund and Copenhagen universities. The strategic role of universities as the key providers of new knowledge is evident with respect to promoting the formation of biotechnology and other high-tech clusters. Lund University has undergone a transition from a traditional Humbolt type university to become an entrepreneurial university taking and implementing strategic decisions (Melander, 2006). One example of such strategic decision making is the building of transdisciplinary and transfaculty research centres, such as the Biomedical and Stem cell centres mentioned earlier in the report, which is located in the so-called tenth area directly under the vice-chancellor.

In Sweden, as is the case also in Finland and Norway, the universities have got a so-called "third mission" in addition to teaching and research, *i.e.* to co-operate with the surrounding society on everything from commercialisation of new knowledge to policy advice. Denmark on the other hand tries to achieve this by giving the universities an external majority in their boards, which elects the vice-chancellor, and by giving increased authority to appointed leaders on faculty and departments levels. Another part of this strategy is the initiative to force universities to merge with other universities and research organisations in order to get bigger and stronger universities. The merger of Copenhagen University with Danish University of Pharmaceutical Science and Royal Veterinary and Agricultural University mentioned before is one such example, becoming one of three so-called "super universities" in Denmark. This merger might have some positive effects on strengthening basic research relevant for the biotech industry, and, thus, for Medicon Valley.

Moreover, another important element of the bottom-up initiative behind the establishment and development of Medicon Valley is the efficient and well-functioning public-private partnerships with respect to research collaboration between university and both, Big Pharma and small DBFs, venture capital raising, and general support for cluster formation. The regional level offers particularly favourable conditions for such partnerships due to the presence of social capital as well as spatial and social proximity between various actors and agencies.

Future policy challenges

Based on the above analysis, the following strengths, weaknesses, opportunities, and threats for Medicon Valley are identified (see Table 5.3).

Table 5.3. **SWOT analysis of Medicon Valley**

Strengths	Weaknesses
• Local presence of Big Pharma • Large number of small and medium sized DBF • Strong academic research • Extensive breadth in value chain components • World leading in four therapeutic strongholds • Presence of network organisation Medicon Valley Alliance • Embedded in global knowledge networks	• Cluster fragmentation along national borders • Imbalance • Weak commercialisation of academic research
Opportunities	**Threats**
• Inter-sector collaboration (food) • Increased cross-border integration • Quality of life	• Dependence on Big Pharma

In principle Medicon Valley has many essential cluster components in place in terms of key players in the drug development value chain (big pharmaceutical companies co-operating with small and medium sized DBF in new drug development), support infrastructures and presence of skilled researchers. As such, it ranks high in the hierarchy of global bioregions. However, the cluster is unique in being located in a region that spans parts of two different countries. This feature can and should not be ignored as a future policy challenge of the cluster compared to similar life science clusters in scope and size such as nearby Stockholm-Uppsala. The national border contains a liability for fragmentation of cluster activities. The full potential for synergy effects derived from close proximity may be hampered because of this. In other words, cross-border regional interaction is only weakly developed. In addition, extensive parts of the regulative environment (*e.g.* tax-rules, employment legislation) as well as research and innovation policy is shaped and implemented within a national framework which complicates cross-border cluster interaction.

Diversification opportunities lie primarily in exploiting biotechnology as a generic platform technology through expanding its application into new areas of related variety, *e.g.* green and white biotechnology. This diversification process has already begun with the development of bio-agro research and industry connected to the VINNOVA Vinnväxt initiative "Food innovation at interfaces", which as one of its potential growth areas focus on "functional food". Building on the idea of related variety will secure maximal knowledge spillover effects by combining industries with complementary and differentiated knowledge-bases (Asheim *et al.*, 2006).

Lessons for other clusters

From earlier research (Asheim *et al.*, 2003a) the need for a more system-oriented as well as a more proactive innovation based regional policy has been emphasised. A re-orientation of what is called "the target level of support", changing innovation policies from being firm-oriented to "a (regional) system-oriented" perspective has already gained growing attention among researchers and policy makers (Asheim *et al.*, 2003b). However, the second part of the recommendation concerning the "form and focus of support" implying a change of focus from allocation of resources for innovation to focusing on "learning aiming for behavioral value-added" has only been addressed more recently through new policy initiatives by EU and various national and regional agencies of "constructing regional advantage" (Asheim *et al.*, 2006).

The theory of constructed advantage allows for more attention to the role and impact of the public sector and policy support, preferably in public-private partnerships, by acknowledging to a greater extent the importance of institutional and economic complementarities in knowledge economies than theories of comparative and competitive advantage do. Institutional specificities constitute the context within which different organisational forms with different mechanisms for learning, knowledge accumulation and knowledge appropriation evolve. Instead of market failure, the rationale for policy intervention is the reduction of interaction or connectivity deficits which lies at the core of a networked regional innovation systems approach.

Innovation policy tools can be classified in two dimensions, resulting in a four quadrants table (see Table 5.4). The recommendations of pursuing innovation support policies that are more "system-oriented" and more focused on "learning aiming for behavioural value-added" are, thus, clearly confirmed. The key role played by the network organisation Medicon Valley Academy in promoting collaboration and networking in the cluster is very

Table 5.4. **Two-dimensional classification of main innovation policy instruments**

	Support: financial and technical	Behavioural change: learning to innovate
Firm focused	• Financial support • Brokers	• Mobility schemes
System focused	• Technology centre	• Regional innovation systems

Source: Asheim *et al.*, 2003.

illustrating of the importance of a system-oriented approach, while the weaknesses in commercialisation of academic research is a telling example of the need for a change in the behaviour of all actors in a "triple helix" context.

The SWOT analysis emphasised furthermore the competitive advantage of having a combination of Big Pharma and a stock of small and medium-sized DBFs, the presence of strong academic research as well as well-developed links to the leading global nodes of the industry. These factors point at the necessity of both building-up and strengthening the endogenous knowledge infrastructure (universities and research institutes) and stimulating co-operation with national and international leading research institutions and companies. The Nordic tradition for co-operation and collaboration, also found in Medicon Valley, is according to the SWOT analysis perhaps the most important individual factor contributing to its success. Of these collaborative relationships university-industry co-operation has been by far the most important and successful, while the public sector's contribution has been of minor importance, and can partly be said to be responsible for some of the shortcomings of the cluster (*e.g.* the lack of harmonisation in policies between the two parts of the cluster which, however, must be blamed on the two parts belonging to different national states).

The encouraging lesson to be learned from Medicon Valley is, however, how much can be achieved and accomplished on a regional level, if the necessary and strategic resources as well as the will, capacity and ability to co-operate are present and utilised.

List of abbreviations

DBF Dedicated Biotechnology Firm

DK Denmark

MV Medicon Valley

MVA Medicon Valley Academy

R&D Research and Development

SE Sweden

VC Venture Capital

References

Asheim, B. *et al.* (2006), *Constructing Regional Advantage: Principles, Perspectives, and Policies*, European Commission, DG Research Report, Brussels.

Asheim, B., A. Isaksen, C. Neuwelaens, F. Tödtling (eds.) (2003a), *Regional Innovation Policy for Small-Medium Enterprises*, Edward Elgar, Cheltenham.

Asheim, B., L. Coenen, M. Svensson-Henning (2003b), "Nordic SMEs and Regional Innovation Systems – Final Report", Nordic Industrial Fund, Oslo, available at *www.nordicinnovation.net*.

Asheim, B. and M. Gertler (2005), "The Geography of Innovation: Regional Innovation Systems", in J. Fagerberg, D. Mowery and R. Nelson (eds.), *The Oxford Handbook of Innovation*, Oxford University Press, Oxford, pp. 291-317.

BCG (2002), *Commercial Attractiveness of Biomedical R&D in Medicon Valley. The Role of R&D in Attracting Regional Investments*, Boston Consulting Group, Copenhagen.

BMC (2006), BMC – biomedicinskt centrum, available at *www.med.lu.se/bmc*.

Bloch, C. (2004), "Biotechnology in Denmark: A Preliminary Report", Working Paper 2004/1, The Danish Centre for Studies in Research and Research Policy, University of Århus, Århus.

Cooke, P. (2005), "The Evolution of Biotechnology in Bioregions and their Globalisation", paper prepared for the "Unlocking Biotechnology" sub-theme of the EGOS Colloquium, 30 June-2 July, 2005, Berlin.

Melander, F. (2006), *Lokal Forskningspolitik. Institutionell dynamic och organisatorisk omvandling vid Lunds universitet* 1980-2005, Lund Political Studies, Department of Political Science, Lund University.

Moodysson, J. (2007), *Sites and Modes of Knowledge Creation: On the Spatial Organization of Biotechnology Innovation*, Meddelanden från Lunds Universitets Geografiska Institution, Avhandlingar CLXXIV, Department of Social and Economic Geography, Lund University.

Moodysson, J. and O. Jonsson (2007), *Knowledge Collaboration and Proximity: the Spatial Organisation of Biotech Innovation Projects*, European Urban and Regional Studies, Vol. 14, No. 2.

MVA (2001), "Medicon Valley Investment Guide 2001", Medicon Valley Academy, Lund and Copenhagen.

MVA (2004), "From Bioscience to New Jobs in Medicon Valley: A Medicon Valley Academy Strategic Report", Medicon Valley Academy, Lund and Copenhagen.

MVA (2005), "Medicon Valley Annual Report 2004", Medicon Valley Academy, Lund and Copenhagen.

MVA (2006), "Medicon Valley Annual Report 2005", Medicon Valley Academy, Lund and Copenhagen.

Vinnova (The Swedish Governmental Agency for Innovation Systems) (2001), "The Swedish Biotechnology Innovation System", Vinnova (Swedish Agency for Innovation Systems), Stockholm.

Vinnova (2005a), *Nationella och regionala klusterprofiler. Företag inom bioteknik, läkemedel och medicinsk teknik i Sverige* 2004, Kompletterings-PM 05 05 05 med hela Medicon Valley, Vinnova (Swedish Agency for Innovation Systems), Stockholm.

Vinnova (2005b), *Strategi för tillväxt – Bioteknik, en livsviktig industri i Sverige*, Vinnova, Stockholm.

Ørestat (2007), Open Øresund Data Bank, *www.orestat.scb.se/website/index.aspx*.

CLUSTERS, INNOVATION AND ENTREPRENEURSHIP – ISBN 978-92-64-04442-5 – © OECD 2009

ISBN 978-92-64-04442-5
Clusters, Innovation and Entrepreneurship

Chapter 6

The Engineering Cluster of Dunedin, New Zealand

by
Ifor Ffowcs-Williams,
CEO, Cluster Navigators Ltd, Nelson, New Zealand

This chapter illustrates how a remote medium-sized city can transform from losing human capital and industry, to becoming one of the most important economic centres of its country, on the edge of engagement with the world economy. This case will show that communication is important to attract investment and highly qualified talents to a cluster in a less known region. It will also demonstrate that active policies and initiatives are essential to ensure the availability of human capital in the cluster. Moreover, this chapter addresses the challenge of building networks of trust and collaboration in an environment in constant change due to important flows of migration. Finally, Dunedin is also a good example of the efforts required to build and strengthen international linkages in order to expand the market and increase the network of partners and suppliers abroad.

Introduction

In the late 19th century Dunedin was New Zealand's main industrial and commercial centre, servicing the rapidly growing gold mining, shipping and rail industries. During these prosperous times many institutions and businesses were established in Dunedin: the colony's first daily newspaper, its first university, art school and medical school among them.

With the passing of the gold rush, the more rapid growth of cities to the north of Dunedin, and the major New Zealand recession of the 1980's, the relative importance of Dunedin was severely reduced. Between 1976 and 1985 Dunedin lost 10 000 people; between 1987 and 1991 over 5 600 jobs were lost. Dunedin is now New Zealand's fifth largest town with a population of 120 000.

Dunedin has been described as "The most southerly city in New Zealand ... one of the most remote economic centres on the planet .... with a cluster of engineering firms that are on the edge of engagement with the world economy" (Campbell-Hunt, Chetty and Matear, 2005).[1]

Many other engineering clusters around the world have succeeded in developing specific and narrow capabilities leading to the competitiveness to profitably service distant markets. For example:

- Taranaki, New Zealand[2] with an oil and gas engineering focus.
- Sweden's Robot Valley.[3]
- Marine engineering, Saint Nazaire.[4]

Dunedin has yet to develop a similar core competency.

Nature and evolution of the cluster

As the surviving engineering firms slowly emerged from the period of rationalisation in the 1980's, it became evident the industry still had a number of problems to address:

- The rapid removal of tariff barriers was opening up the domestic market to a flood of imports.
- A high New Zealand dollar was having a significant impact on export prospects, particularly the SMEs.
- As firms battled to survive, apprenticeship programmes were largely overlooked and this impacted on school leavers who saw engineering as a poor career choice.

 CLUSTERS, INNOVATION AND ENTREPRENEURSHIP – ISBN 978-92-64-04442-5 – © OECD 2009

- Large international orders were being lost by the lead firms because of the lack of suitably qualified staff and the absence of an apprentice intake.
- Old rivalries were still evident; competition was fierce, with little trust existing between firms. This limited familiarity impeded the ability of the firms to deal with common industry problems and to develop complementary areas of expertise... co-specialisation and out sourcing.
- There was a perceived view that Dunedin engineering firms were servicing a small local market, with low growth, and offering little opportunity.

The accumulated pool of engineering experience within Dunedin was undermined during the severe 1980's recession. More recent employment data is showing a significant turnaround: the number of full time employees increased by 13 per cent in 2004 to 2 334 at a time when employment in the region grew by only 2 per cent and national employment in engineering increased by only 3 per cent.

Today Dunedin is home to a broad base of engineering firms producing a range of short-run complex products. Within the cluster a number of firms have been experiencing significant growth. Total revenues are now around NZD 200 million (approximately EUR 100 million).

There are around 80 engineering firms in total, of various sizes and capabilities, with eight of the more export intensive firms having over 50 staff each. Two sub-groups within the cluster are evolving:

1. The rainmakers

   These firms are servicing established international markets with niche, design-to-order products. They are strong on design intensive products and innovation. There is co-specialisation between these firms, their product ranges do not compete and they are proactive in developing partnership arrangements with local suppliers. They network and collaborate nationally and internationally; their key relationships tend to be outside of the city. They are able to accumulate and internalise learning. Employment is typically between 50 and 100 or more.

2. The feeders

   These smaller firms focus on the local market; many are satisfied with their current level of activity. Their product range is typically broader than the rainmakers. Many are jobbing shops servicing the local market. There is increasing collaboration and co-specialisation between these firms. Accumulated learning tends to be shared in the community. Employment is often around a dozen staff.

A snapshot of five Dunedin rainmakers

DC Ross: a Dunedin tool-maker and precision engineering firm that exports car components. Each year, DC Ross produces about 10 million individual automatic transmission, brake, and seat and plumbing parts for cars as diverse as Ford, Holden and Maserati. Australia accounts for 75 per cent of its business, thanks in part to a direct shipping route from Dunedin to Melbourne. The firm has around 28 staff.

Farra Engineering: one of Dunedin's oldest engineering companies with roots going back 140 years. The Managing Director is a fifth generation member of the family business that began looking for off shore work in the 1980's, focusing on its access hoists for the external maintenance of high rise buildings. These are now exported to Singapore, Hong Kong, and UK, and generate spin-off work for local sub-contractors in hydraulics and electrical controls. "Offshore we are regarded as the company that can do difficult work in a short time"[5] Engineering Dunedin Inc., the cluster's association, is currently chaired by Farra's Managing Director. The Farra Engineering group has around 130 staff.

Fisher and Paykel (F&P)**:** major manufacturer of white goods, especially dishwashers. The firm was initially an importer, then in the 1930's started manufacturing under license, and finally manufacturing to its own designs which have become world firsts. The firm now also design and build their own production lines, and production lines for other appliance manufacturers around the world.

In the early 1990s a young designer at the Dunedin factory looked sideways at his filing cabinet and thought for one crazy moment that it might make a dishwasher, with drawers. The DishDrawer, as his idea became, took eight years to develop and is now one of F&Ps leading exports. This manufacturer has approximately 680 staff.

Millers Mechanical: this group produces specialised engineering equipment for abattoirs, exporting to Australia, Japan and USA. It has around 74 staff.

Scott Automation Systems: designer and manufacturer of automated process and production equipment. The parent company was established in Dunedin in 1913. The firm's association with the appliance industry dates back to the 1940's when Scott Washing Machines and Refrigerators were manufactured under license from Whirlpool and Norge. The parent firm is a New Zealand-owned, publicly-held engineering company. The company is recognised by all the major appliance builders in the USA and internationally as a world class builder of advanced automation systems. Scott is currently working on robotic systems for the meat industry, a world first. This firm has approximately 60 staff.

 CLUSTERS, INNOVATION AND ENTREPRENEURSHIP – ISBN 978-92-64-04442-5 – © OECD 2009

Development of the clustering intervention

The clustering initiative started in Dunedin in 1998 with the encouragement of New Zealand's national export organisation. This led in 2003 to the establishment of a fully Incorporated Society known as "Dunedin Engineering Inc.".

The Dunedin City Council (DCC) Economic Development Team has been instrumental in supporting Engineering Dunedin Inc. The cluster's leadership team includes the CEO's of the major engineering companies. DCC facilitates the Leadership meetings and full cluster meetings.

Financial support from DCC, supported with small grants from the national agency, has enabled a wide range of projects and activities to be developed, including:

- Building the social connections amongst the diverse range of "competitors". DCC state that this simplest of support programmes (soft networking) took at least five years to show any significant return to the cluster or the local economy. This underlines the need for patience when trying to expedite the development of what is essentially an organic entity.
- Early initiatives involved development of database/directory of competencies which was made available in hard copy and on the website and promotional material for members to take offshore. This proved quite effective when trying to secure large international orders, especially for F&P DishDrawer in US. Members were able to demonstrate that they were from a location with some critical mass in engineering, earning credibility by association.
- Promotion of Engineering within the region as a career option for school leavers, and the development of Engineering Career Days.
- Promoting Dunedin's engineering employment opportunities in conjunction with recruitment agencies at Expos; seeking skilled migrants with the UK Immigration Expo (one company has employed 25 new immigrants).
- Engineering Dunedin Inc. and the local Otago Museum launched a Robotics Competition for Otago school students, drawing together engineering firms and education providers.
- Forwarding appropriate CV's on to firms seeking skilled employees.
- Feasibility study regarding a Major Regional Design Initiative.
- Development of the Engineering Dunedin website.[6]
- The development of the "Supply Cluster" at Fisher and Paykel.
- Seeking to fill gaps in the local cluster's capabilities, *e.g.* attracting a hot dip galvanising firm to Dunedin.

- Close links with the local polytechnic regarding research, student employment opportunities and international business to business linkages. A Reciprocal Student Exchange to and from Ulm, Germany is in place.
- Assistance regarding the successful joint application with the Otago Polytechnic for "Polytechnic Regional Development Funding" of NZD 257 000. A full audit of Dunedin engineering companies will be carried out regarding current skills gaps, new apprentice skill requirements and possible synergies regarding capabilities that may exist between firms. The Polytechnic will then work with engineering firms to deliver appropriate courses, relative to business needs either on a firm's site or at the Polytechnic.
- Media enquiries and presenting to the media appropriate engineering stories.
- Arranging regular meetings with guest speakers to draw together the cluster's firms.
- Securing funding from the national agency to co-fund cluster joint activities.
- The EDU has also drafted a Code of Ethics and Incorporation document for the Engineering Cluster, and funded the legalities associated with Incorporation.

Success factors

- The cluster originally emerged in the 19th century in response to a specific local demand to service developing industries, in particular foundries for the gold mining activity inland of Dunedin and engineering for the sea and rail transport sectors.
- There are indications that firms within the cluster have a higher ratio of R&D investment to sales than is typical of New Zealand firms in general.
- The cluster's firms also have a higher ratio of investment in process equipment and in training and education (Greatbanks, Batley and Everett, 2006).
- Sensitivity to exchange rates, especially the American dollar.
- Stable workforce.
- The availability within the cluster of a wide range of subcontractors offering increasingly specialised services to the lead exporting firms.
- Flexible support.
- Co-specialisation, consolidating of demand for components and sub-assemblies.

Role of SMEs

- SMEs are a key to the competitiveness of the Dunedin cluster, providing the lead exporting firms with a wide range of competencies that they do not need to maintain in-house.

CLUSTERS, INNOVATION AND ENTREPRENEURSHIP – ISBN 978-92-64-04442-5 – © OECD 2009

- A number of Dunedin firms through their cluster involvement are now utilising other cluster members to carry out work for them, whereas firms from outside the province previously carried out this work.
- Whilst new firms are emerging, conservative traditions within the community limit entrepreneurial activity.

Impact of the cluster on entrepreneurship and employment

- New firms, some started with redundancy funds, are emerging in Dunedin because of the reputation for engineering excellence amongst the larger firms (Brown, 2005).
- Firms are now able to tender for larger contracts than was feasible before the start of the clustering engagement.
- Collaboration in an array of forms is leading to an upgrading in competitiveness and employment growth. Collaboration now includes the joint promotion of engineering education; joint commitment to train apprentices, capacity-sharing and staff sharing to even out spikes in demand in this lumpy, contract-driven industry; and subcontracting components.
- Companies are now helping each other out in unprecedented ways ... bringing their skills together on new products ... borrowing staff from each other... subcontracting out more to cope with increased book orders.

Three hard business networks have emerged from the clustering initiative. These are consortiums/strategic alliances between smaller groups of firms, frequently SMEs:

1. **United Tooling Solutions** (UTS) was formed in 2001 with help and funding from the Dunedin City Council's economic development unit. Its objective was to market, co-ordinate and work for three precision tool manufacturers within the engineering cluster. This network came out of high level discussions; Fisher and Paykel (F&P) were having tools made outside of Dunedin, and in order to secure F&P work the toolmakers had to secure other work to balance out the peaks and troughs. Several million dollars in orders were gained by UTS' work, but in 2006 UTS was voluntarily wound up after nearly five years operation (Otago Daily Times, 2006). An extended period of a high New Zealand dollar and an influx of cheap China-made tools during the previous 18-24 months meant UTS was no longer viable for its three joint venture companies.
2. **Dunedin Marine Construction:** Dunedin has a solid ship building past which today is underutilised. Drawing on this accumulated competency, 20 firms (some from beyond the local engineering cluster) were identified through a capability assessment funded by a national agency, Technology New Zealand. Subsequently these firms contributed NZD 100 000 between

them (with matching funds from Industry New Zealand, another national agency) to seek tenders for the maintenance of fishing vessels and tourist launches. Although work has not been steady, projects have been completed on time, to budget and to a high standard. A tender is currently being prepared which if successful would involve input from 15 firms over a 12 month period.

3. **Foundry group:** As a result of the informal soft networking through cluster meetings three foundries began to work together in prompting their complementary skills internationally. Following a visit to Australia, an initial NZD 140 000 order was secured. Joint initiatives have not occurred over the last 12 months due to each of the independent foundries being kept busy with their own projects.

There has been a steady increase in employment since 2001 as shown in Figure 6.1.

Figure 6.1. **Full time equivalent employment trend**

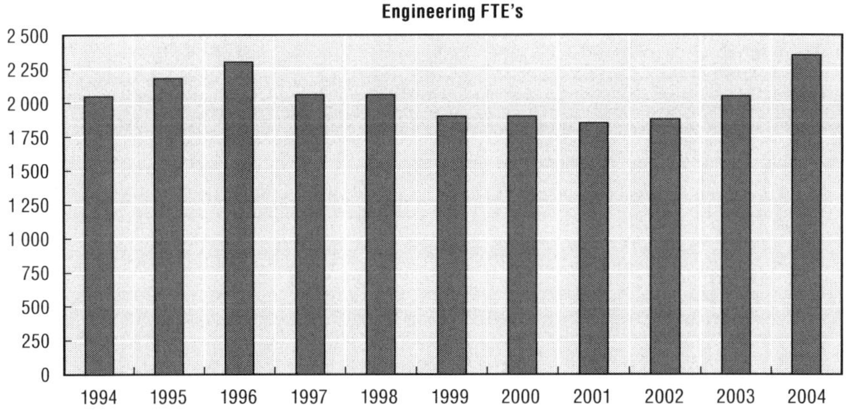

Source: Dunedin City Council Report, 4 October 2005.

Barriers to cluster development

- While Dunedin has (for a young country) a strong and well-established engineering tradition, the cluster remains vulnerable to the possibility, however remote this may be, of one of the major rainmaking firms exiting.
- A current weakness in the industry is attracting the "right" young people into engineering as a career option. This problem relates partially to the industry contracting in the 1980's and early 1990's and to parents, teachers and young people's impressions of a "dirty industry". There is a need to continue boosting the profile of the industry with young people.

 CLUSTERS, INNOVATION AND ENTREPRENEURSHIP – ISBN 978-92-64-04442-5 – © OECD 2009

- Dunedin engineering also needs to continue working on skilled staffing issues as well as possible joint purchasing initiatives.
- Dunedin engineering firms in the future will need to foster ongoing innovation and product development.
- Opportunities exist to improve the interface between educational institutes and the engineering sector. The local university has yet to substantially engage with the cluster.
- Other opportunities may emerge regarding co-specialisation, which will hopefully foster innovation and ongoing product development.
- Continuing to move away from a jobbing approach to specialisation.
- Continue moving from passive co-location to active networking and co-operation.
- Exchange rate fluctuations.

Role of policy

- In 1996 the New Zealand Trade Development Board, a national agency, introduced the concept of clusters to regional development agencies throughout New Zealand with a series of regional presentations.
- Dunedin was an early mover in New Zealand in introducing a clustering approach, with industry clusters becoming the basic unit of focus for the Dunedin economy from 1997. The emphasis moved from investment attraction to supporting the city's own entrepreneurs, the "grow your own" programme as it was known. The small budgets were initially used to encourage soft networking of firms within the cluster in an effort to build trust and to stimulate joint commercial activity in the form of sub-contracting. Soon after soft networking began, research on cluster capabilities and market opportunities was undertaken.
- DCC's direct financial support in 2004/5 for the cluster amounted to NZD 35 000. In addition was a part time cluster facilitator, employed by DCC. The staff member appointed was a mature, seasoned professional with strong networking skills. This person has been in the role for 4 years. He has been a major player in developing the cluster's social capital, and driving key projects, with strong support from the senior members of DCC's economic development team. He is proactive in encouraging the Executive Members of the cluster to take responsibility for a number of activities and projects.
- The hardest thing for local government involvement/investment in clustering interventions is the desire of politicians to see immediate results. As a minimum, a five year time horizon is needed, well beyond the usual

political cycle. In Dunedin it was necessary to fight to keep the resources for this project because of its long-term nature and the difficulty in showing a causal link.

- One of the cluster's first projects was to develop a mechanism to share apprentices between firms, so that young employees were guaranteed work within the cluster without it necessarily being with just one firm. Today even skilled staff are shared between firms; one firm has had staff from 26 other firms working on export orders during high demand periods.
- The driving force behind turning the cluster into a more dynamic actor within the local economy was the development of trust between firms that led to the harnessing of existing technical capabilities to target new markets. The firms needed a neutral corner to explore each other's capabilities without risk to themselves. The clustering initiative, resourced largely by local government, provided that, and led to firms choosing to work together with other firms they trusted.

Policy adaptations over time

Central and local government have had some major impacts on the development of the cluster.

On the positive side:

- A national agency introducing the merits of engaging locally through a clustering initiative, and offered small scale financial support for cluster facilitation during the initial stages.
- The local government stimulating the development of social networks, the cluster group meetings, bringing together firms in a neutral setting.
- A noticeable change in attitude… the cluster no longer waiting for national initiatives, support.
- Collaborative offshore marketing programmes co-funded by a national agency.
- Programmes of skill development in the region's labour market.
- Recently a national agency has sought to improve linkages between engineering clusters within New Zealand, in particular Taranaki and Dunedin.

And on the negative side:

- Radical reform of apprenticeship training.
- Radical economic reform leading to the major recession in the 1980s.
- Difficulty in obtaining tight alignment between the needs of the local firms (with Dunedin City Council responding well to those needs) and the more "top down" approach of the national agencies.

Future policy challenges

- Arguably, this small remote cluster in a far corner of the globe has yet to establish sustainable critical mass.
- There is vulnerability at this stage of the development of the cluster that rainmaker firms move resources out of Dunedin, possibly on a large scale. The key relationships (and ownerships) of many of the rainmakers are outside of Dunedin.
- Attracting school levers, and retaining within Dunedin the skilled workforce.
- The Polytechnic has been proactive in collaborating with the cluster; the University (aside from the Department of Management) has yet to engage in a meaningful way.
- Most of the engineering businesses ... the feeders ... are comfortable with their current size, take a low-risk approach and are reluctant to significantly invest in new facilities.
- Continuing the move from jobbing (selling machine time) to own product, with own IP.
- Once fierce competitors have over recent years learnt to work together, but there is still scope to opening up further the trust and dialogue to accelerate the move to co-specialisation.
- Blowing its own trumpet... nobody is going to do it for them!

Lessons for other clusters

- Importance of taking a long-term perspective, with (long-term) public funding being used to continually drive the agenda forward in upgrading the competitiveness of the cluster.
- Sustained and steady intervention will be more effective:
  - ❖ Steady investment in the region's skill and knowledge-base, and/or its physical infrastructure.
  - ❖ Sustained policies of support to the processes of regional agglomeration and resource relocation.
  - ❖ And for the development of offshore markets.
- Don't ignore the payoffs provided by a neutral agency in bringing "competing firms" together, building trust and dialogue between them.
- The value of taking time in establishing the legal entity for the cluster group. This should not be an early priority.
- Don't consider an "exit strategy" for the publicly funded agency ... rather consider an exit from maturing projects/initiatives which should over time become "owned" and then financed by the benefiting firms.

- Garnering whole-of-government' support for a cluster does not come easy. Aligning different government agencies at national and local levels takes time and energy. Cluster development is a local activity, yet national agencies have difficulty in empowering those closest to the cluster in the allocation of national funds.
- Don't wait for national support/hand outs ... consider these as an extra, a bonus rather than essential for the life of the clustering initiative.
- Value of resourcing long-term a senior, mature professional with strong networking skills as the cluster's facilitator.

Acknowledgements

Grateful thanks are due to Des Adamson and Peter Harris from Dunedin City Council's Economic Development Unit. Also to Damian O'Neill and Peter Brown, who were instrumental in establishing this clustering initiative when they were with Dunedin City Council. Professor Colin Campbell-Hunt of Otago University has also contributed significantly to this chapter.

Notes

1. This review draws extensively on this paper and on reports generously provided by Dunedin City Council's Economic Development Unit.
2. See *www.engineeringtaranaki.co.nz*.
3. See *www.robotdalen.org/english/english.htm*.
4. See *www.pole-marine-atlantique.com*.
5. Rod Oram quoting the then General Manager.
6. See *www.cityofdunedin.com/city/?page=cluster_engineering*.

References

Brown, Peter (2005), "Walking the Talk – Lessons from Dunedin's Cluster Development Programme", Dunedin City Council.

Campbell-Hunt, Chetty and Matear (2005), "Clustering at the Edge, Growing Businesses of Global Reach from Thin Local Soil" (Working Paper in Competitiveness), University of Otago.

Greatbanks, Batley and Everett (2006), "International Manufacturing Strategy Survey", University of Otago, Department of Management, PowerPoint presentation 29 May 2006.

Otago Daily Times, Dunedin 24 July 2006, p. 18.

Rod Oram (2002), "Pride of the South", *Unlimited Magazine*, April, Infego Communications Ltd.

ISBN 978-92-64-04442-5
Clusters, Innovation and Entrepreneurship

Chapter 7

The University-centric High-tech Cluster of Madison, United States

by
Martin Kenney, Amanda Nelson and Donald Patton
University of California, Davis

This chapter shows the central role played by the university in promoting economic development, innovation and knowledge across the region. The initiatives of the national and regional governments to spread the outcomes of the university to the regional economy are well illustrated in this chapter. The case of Madison, Wisconsin, also illustrates the various efforts made by the university to encourage commercialisation, licensing and technology transfer from the university to industry. With the support of related bodies such as the alumni association, the faculties and the technology transfer bureau, the University of Wisconsin Madison shows that universities can also play an important role in linking innovations to venture capitalists and industry, in stimulating the creation of spin-offs, and in facilitating the identification of the market for new products.

Introduction

The city of Madison, Wisconsin covers 111 km^2 in mid-western United States and is surrounded by farmlands and rolling hills. Within this farmland sits a hotbed of high-technology activity. Home to a flourishing biotechnology industry, the city of Madison is particularly interesting as a focus of high-technology cluster research as it is a location that has undergone a reinvention in the last two decades. In 1980, the city was dependent upon public sector employment due to its position as Wisconsin's state capital and the presence of the state's flagship public university, the University of Wisconsin, Madison (UWM) – an educational institution with a large student base and significant direct employment. The private sector was composed of services, smaller manufacturing plants, and significant industrial food processing, in particular meat processor Oscar Meyer. The manufacturing and food processing portions of the Madison economy have collapsed. Despite this loss the public sector would certainly have been sufficient to prevent too precipitous an economic decline, though Wisconsin itself has experienced economic difficulties.

In roughly 1980, UWM began to experience an upswing of entrepreneurship from its faculty – a development that is roughly contemporaneous with the commercialisation of university molecular biological research (Kenney, 1986). This trend has accelerated in recent years. While Madison area employment in manufacturing has remained essentially static from 1998 through 2004 (3.6 per cent growth over six years), employment growth in the life sciences has been 54.4 per cent over this same time period. Similarly, employment growth in information technology (44.3 per cent) and physical sciences (22.6 per cent) between 1998 and 2004 indicates that Madison is increasingly dependent upon high-technology enterprises to provide employment opportunities. (See Figure 7.1 and Table 7.1 on Dane County, Wisconsin [Madison area] employment.)

In 2007, Madison had a vibrant entrepreneurial cluster that was motivated by the knowledge and innovation of the UWM faculty, students, and local alumni. We report on the genesis and evolution of this cluster using primary source material on the origin of the entrepreneurial start-ups, various secondary sources of aggregated statistics, and the results of personal interviews.

Figure 7.1. **Dane County employment, 1993-2004**

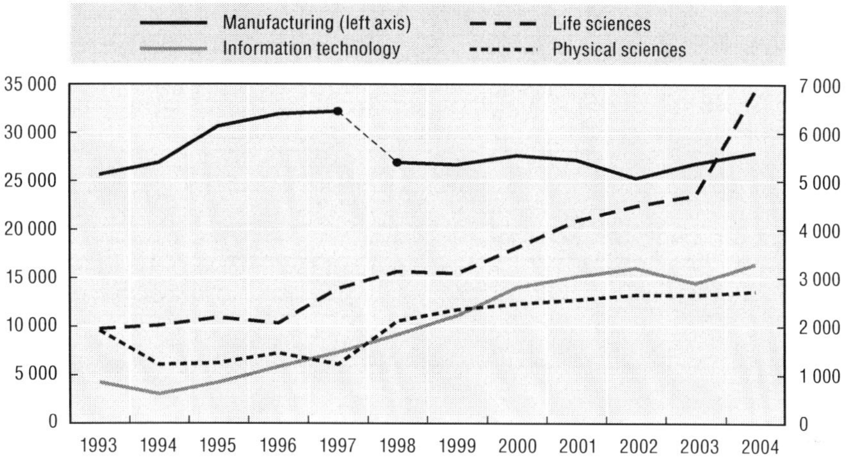

Source: County Business Patterns Data, US Census Bureau.

Table 7.1. **Dane County employment**

	Life sciences	IT	Physical sciences	Manufacturing
1993	1 952	849	1 916	25 652
1994	2 022	612	1 218	26 896
1995	2 181	847	1 250	30 640
1996	2 066	1 172	1 453	31 916
1997	2 777	1 468	1 220	32 192
1998	3 138	1 834	2 107	26 902
1999	3 099	2 224	2 345	26 681
2000	3 630	2 801	2 463	27 623
2001	4 183	3 031	2 546	27 166
2002	4 489	3 214	2 653	25 299
2003	4 707	2 897	2 646	26 823
2004	6 883	3 294	2 722	27 920
Growth rates (%)				
1993-2004	71.60	74.20	29.60	
1993-1997	29.70	42.20	–57.00	20.30
1998-2004	54.40	44.30	22.60	3.60

Source: County Business Patterns Data, US Census Bureau.

Nature and evolution of the cluster

Madison, located in southern Wisconsin, has a population of just over 220 000 persons and has one of the highest concentrations of advanced degrees in the US at 2 per cent of the residents. Actually identifying the number of "high-technology" firms in Madison depends upon definition. For example, the broadest definition, which includes small PC assembly and routine blood testing laboratories, is typical as many jurisdictions in the US and around the world inflate their number of high-technology firms and entrepreneurial start-ups. Therefore some sources claim that there are currently between 450 and 500 high-tech firms in Madison – this would be a growth of approximately 14 per cent over the last five years (Ladwig, 2004; MG&E, 2006). If all of these firms are taking together, they provide greater than 8 per cent of the region's total employment or 26 000 jobs (MG&E, 2006). The statistics on biotechnology firms are even more discrepant. For example, Willis (2004) claims there are approximately 250 biotech firms in the region that have revenues of nearly USD 5 billion annually. The Madison Gas and Electric (M&GE, 2006) claims there are 112. Our rigorous count of only the entrepreneurial biotechnology firms that excludes blood testing laboratories, hospitals, seed testing laboratory and the like finds there are 59 such firms, 33 of which are direct UWM spin-offs by founder. Clearly, definitions are all important.

Madison is one of the most politically liberal (in the social democratic sense, not traditional English liberalism) cities in the Midwest, and yet it also is considered business friendly. This has been recognised in the US media. For example, in 2004 Madison was ranked the number one US city for business by Forbes, a well-known US business journal (Tatge, 2004). A senior editor of the magazine justified this in the following way, "Madison's number one ranking is a result of its labour supply, strong income growth, as well as the fact that the city ranked tops in per capita number of PhD's and third highest in the US in terms of people with college degrees" (MG&E brochure, no date). In addition to the economic success of the region, it attracts residents with a vibrant downtown area and multitude of recreational opportunities, including boating, biking, skiing, art and entertainment.

Madison is a knowledge cluster, based largely on research, in particular, in biotechnology done at the UWM. Of the 182 high-technology firms identified by our research, just under half have direct ties to founders from UWM. The extent to which UWM is a direct source of these firms varies considerably by sector. Over 62 per cent of the 59 biotechnology start-ups in our population are directly related to the university through one of their founders. This level of involvement declines to 50 per cent for biotech support firms, 43 per cent for engineering start-ups, and just one of the 16 IT start-ups has a founder that came directly from UWM. Table 7.2 provides this data in greater detail.

On average, firm size is small with most companies employing 5 to 500 individuals. Early firm creation began in the late 1970s, but the majority of growth in the cluster has occurred in the last 15 years. Figure 7.2 presents new firm formation from before 1960 to the present.

Table 7.2. **Relationship to UWM of the start-up founders**

Industry	UWM founder	No UWM founder	Total
Biotechnology	18	11	29
Biotech Support	15	15	30
Medical	15	8	23
Vet/Ag	4	1	5
Total life sciences	**52**	**35**	**87**
Electronics	9	11	20
Engineering	10	13	23
Telecom	0	2	2
Total physical sciences	**19**	**26**	**45**
IT	1	16	17
Software	11	14	25
Internet	0	3	3
Total information technology	**12**	**33**	**45**
All firms	84	98	182

Source: Author's Database © Martin Kenney and Donald Patton.

Figure 7.2. **New firm formation in Madison, Wisconsin by year**

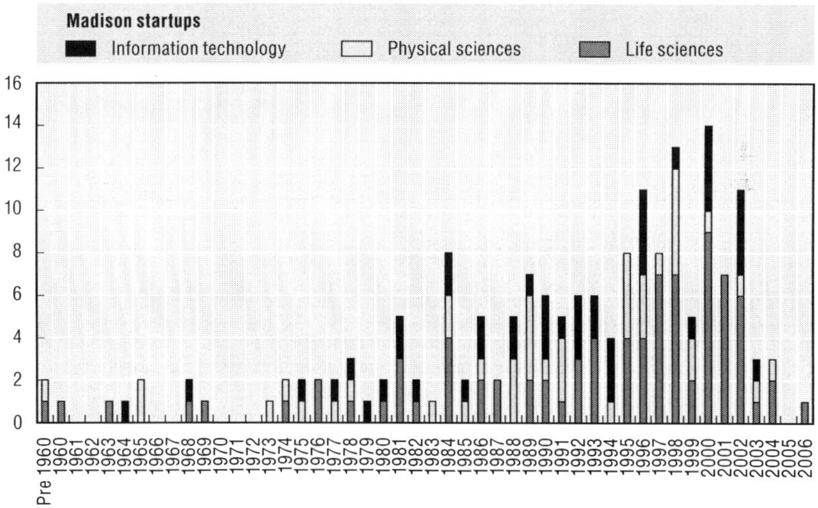

Source: Author's Database © Martin Kenney and Donald Patton.

As a field of entrepreneurship, biotechnology is well suited to the university environment because for the most part new biotechnology firms are direct results of university/medical school laboratory developments whose research was funded by the US National Institutes of Health. Much of this research results in products that a small firm can protect with patents, which if sufficiently promising can attract the interest of angel investors or venture capitalists (for one of the earliest explications of this process, see Kenney 1986).

Though biomedical-related firms make up approximately 50 per cent of the high-technology start-ups in the region, the other 50 per cent are from the physical sciences, engineering, and the IT fields. In these other fields, UWM start-ups are relatively small and have not experienced the growth that IT spin-outs such as SAS from North Carolina State University; Sun Microsystems, Yahoo!, and Google from Stanford; Paypal and Netscape from the University of Illinois, or Cadence and Synopys from UC Berkeley. Having said this, there are a number of high-technology firms in Madison that are not the direct products of UWM, though the university ambiance pervades the city.

Unlike many other high-technology and particularly biotech clusters, around half of the Madison biotech firms are concentrated in the areas of biomedical inputs and services. There are many reasons that can be cited for this focus, but on a basic level, this type of technology involves less risk and less capital. As Randall Willis points out in his 2004 article on Madison, these companies rarely "hit a home run" and accumulate great wealth for the founders or shareholders. However, when successful they can provide consistent and solid returns and good employment opportunities (Willis, 2004). While they may never create enormous capital gains, the start-up costs for input and service firms are relatively low. Harry Burrill from local start-up Lucigen notes that, ... "you can 'bootstrap' starting a products company with much less money, then get products on the market relatively quickly to generate a revenue stream for survival and growth *(ibid)*".

The biotech cluster in Madison does not resemble the "typical" cluster formation based on interaction between similar firms. In reality the cluster is, in large part, a hub-and-spoke morphology with UWM at the centre. UWM dominates the region and is the source of most of the firms. In contrast to Silicon Valley where there have been wave after wave of entrepreneurial spin-outs, in Madison there have been far fewer firms whose founders came from another firm.

The biotechnology research materials firm, Promega, is the notable exception. Promega was founded in 1978 by William Linton. Today, it operates in 11 nations, has 850 employees, has revenues of approximately USD 175 million, and sells 1 450 different life science research materials. Not only is Promega a successful firm, but between 1987 and 2005 three firms have spun-off from Promega. One of the founders of these firms came directly from Promega, while another firm was established with key personnel from Promega. The third firm was PanVera, a company that develops technology used to determine drug components. Former Promega employees founded the company in 1992. Since then six additional companies have spun off of PanVera on the basis of key personnel and founders from the firm.

Although our methodology in establishing firm genealogy is based exclusively on founder(s)' previous employment, a greater understanding of the movement of key personnel allows us to recognise the true importance of these firms to the Madison cluster.[1] Key personnel from PanVera were directly involved in the creation of five firms from 1995 through 2000, although none of them were identified as a founder.

Serial spin-offs are not atypical of the cluster. It has, however, been vital in speeding the expansion of the cluster. Promega was a pioneer and not only have other firms spun-out of Promega and its progeny, but also many employees at other local firms once worked at Promega. So it is a source of entrepreneurs and seasoned executives. As Feldman *et al.* (2005) observe, the early entrepreneurs, if sufficiently successful, actually begin to change their environment. In effect, the environment is not a simple unchanging selection grid, but actually evolves with its resident actors. So, for example, the second-generation entrepreneurs are able to leverage the knowledge and experience gained from initial endeavours in their new firms. The earlier entrepreneurs create awareness and reputation in the community enabling not only them, but next-generation founders to use the pioneers as "proof" that their new concepts have a similar possibility of success thereby allowing them to attract funding and support. By virtue of its success, Promega has been an icon and model for other entrepreneurs in the Madison cluster.

Entrepreneurial support network in Madison[2]

The focus on biomedical inputs and services provides a partial explanation for the lack of venture capital in the region, as these firms do not grow sufficiently quickly and to sufficient size to justify venture capital investment.[3] Conversely, the focus may be a result of this dearth in funding. Although Dane County ranks in the top 100 counties nationwide for venture capital funding, Wisconsin as a whole falls in the bottom half of all states for venture capital investment (Table 7.3).[4] According to PanVera founder Ralph Kauten[rdquoe] there is little venture capital in Wisconsin so most of the start-ups have worked without it[rdquoe] (Kauten, 2006). In 2005 California received the bulk of the nation's venture capital funding (USD 10 220 million or 47.1 per cent of the total). In contrast, Wisconsin received USD 68 million, less than one per cent of the funding (Rosen, 2006).

Venture capital often emerges in entrepreneurial clusters, and many researchers consider venture capital an important indicator of the dynamism of a cluster.[5] Of course, some Madison firms have received venture capital.

Table 7.3. **Venture capital firms operating in Wisconsin**

Firm name	Location(s) in Wisconsin	Date established (date closed)	Reported capital (USD million)	Fund size (USD million)
Advantage Capital Partners #	Madison	1993		
Avolte Venture Fund Midwest # (a sector fund of the Peak Ridge Capital Group)	Madison	2006		
Baird Venture Partners #	Madison and Milwaukee	2001		
Kegonsa Capital Partners, LLC	Madison	1997	11.0	10.7
Madison Capital CIP Corp.	Madison	1982		0.4
Mason Wells Biomedical Fund	Milwaukee	2000 and 2006		
Pangaea Partners	Madison	1989		
State of Wisconsin Investment Board*	Madison	2002	61 500.0	
Venture Investors, LLC	Madison	1982	80.0	9.0
Wisconsin Investment Co.	Madison	1985	1.0	

* Government affiliated.
# Firm has multiple offices, some may be located outside of Wisconsin.
Source: Venture Expert 2007; OCR 2007.

Overall 14 per cent of all Madison start-ups have received venture capital funding. 18 per cent of the life sciences and physical sciences firms have received VC funding, while just 2 of the 45 information technology firms received VC funding (see Table 7.4).

Table 7.4. **Madison start-ups and venture capital funding**

Industry	VC funded	Not VC funded	Total
Biotechnology	8	21	29
Biotech Support	5	25	30
Medical	3	20	23
Vet/Ag	0	5	5
Total Life Sciences	16	71	87
Electronics	2	18	20
Engineering	5	18	23
Telecom	1	1	2
Total Physical Sciences	8	37	45
IT	2	15	17
Software	0	25	25
Internet	0	3	3
Total Information Technology	2	43	45
All Firms	26	156	182

Source: Author's Database © Martin Kenney and Donald Patton.

The funding Madison high-technology start-ups have received reflects this higher interest by VC firms in science-based firms. In the year 2000 information technology firms received just 10 per cent of all VC funding in Madison, compared with life sciences which received 54 per cent of the total. In 2005, after the dot com bubble, life science-based firms increased their share to 80 per cent of all VC funding. Figure 7.3 presents this data in greater detail.

Figure 7.3. **VC backed firms – Dane County Wisconsin**

Venture capital investments, excluding acquisitions, in USD '000s

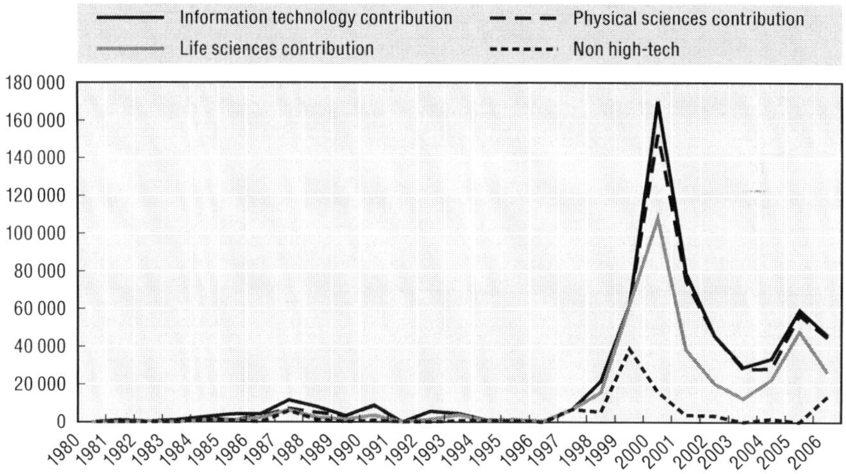

Source: Thomson Venture Economics/National Venture Capital Association, 2007.

Madison differs from clusters such as the San Francisco Bay Area, San Diego, Boston and the Washington DC area because it has successfully created an environment encouraging the formation of new technology businesses without having a concentration of venture capital. To some degree, this absence was mitigated by WARF which was willing to substitute for the venture capitalists in the earliest stages. But the success of Promega, PanVera and others also created an environment within which bootstrapping was understood. In particular, scientists understood how they could use Small Business Innovation Research grants to carry their firms through the early stages of the firm's development. Like traditional VC funding, in lieu of payment WARF frequently accepts equity in a start-up as compensation for their services.

Location

It is not just the business climate in a region that draws entrepreneurs and sparks new firm creation. The programmes and factors discussed above have encouraged and supported the creation of Madison's biotech cluster, but the physical and social environment also plays a role. For Richard Florida,"place has become the central organising unit of our time, taking on many of the functions that used to be played by firms and other organisations," (Florida, 2002: 6). Florida cites diversity and high quality of life as well as a vibrant nightlife and outdoor recreation as important factors that attract and retain individuals.

Though we did not conduct a scientific sampling, there is a general belief in the region that Madison is a desirable location in terms of life style and that the university is an attractive employer. As the founder of Madison company NimbleGen Systems states, "UW is a highly respected research facility and draws talent from across the nation. In the end, the talent really does not want to leave Madison, due in part to the relationships people have formed with a wide range of like-minded people and the attractive standard of living they experience" (Potera, 2004: 2).

Most studies agree that Madison has an excellent public school system, reasonably priced housing, and a wide range of recreational activities. In the winter, skiing is popular, and in the summer Madison offers an extensive bike trail system, golf and hiking. City boosters cite the downtown area as being "vibrant" with the "character of a small city but the amenities of a major metropolitan area" (MG&E, 2006). One widely mentioned drawback is that its relative small size discourages airline connections, thereby making travel more burdensome. For example, one entrepreneur observed that, "there are direct flights to only a very limited number of destinations, and there also seems to be a psychological barrier that hinders companies on the East and West coasts from working with companies in the Midwest" (Willis, 2004). Though logistically it suffers from some handicap, the social amenities appear to provide a level of compensation.

Success factors

This section discusses the institutional drivers of the Madison cluster. The institution at the centre of the Madison high-technology cluster is the UWM. Of all of the technology-based start-ups in Madison, 46 per cent have at least one founder from the university.[6] UWM has a unique institutional model for commercialising research in that the university does not have an office of technology licensing. As a parenthetical note, because UWM is the only major research university in Wisconsin it has increasingly been involved in encouraging entrepreneurship around the state.[7] As the outcome of a historical accident discussed in more detail later, its technology licensing is channelled through an affiliated non-profit organisation called the Wisconsin Alumni Research Foundation (WARF), which is not directly controlled by the university. It is UWM and WARF that have had the greatest impact on the inception and growth of Madison's entrepreneurial ecosystem.

CLUSTERS, INNOVATION AND ENTREPRENEURSHIP – ISBN 978-92-64-04442-5 – © OECD 2009

The university

UWM is at the core of Madison cluster and has been the source of many start-ups. Very importantly, because a university is a decentralised organisation, just because the UWM is at the centre does not mean that the various start-ups have technologies or personnel in common. For example, a software spin-out of the university may share no information transfer paths with a biotechnology start-up. Furthermore it is difficult to establish precisely how many of the start-ups have their origins in the university. In many cases start-ups are direct spin-offs from the university. But in many cases entrepreneurs are emerging from university spin-offs to start yet other firms as is the case with Promega, PanVera and NimbleGen Systems. Of course, these connections are only the most visible and in the discussion of Promega we show that there are many more connections to the point at which it would be fair to say that there is a biomedical materials sub-cluster in Madison that exhibits the "buzz" that geographers have identified with dynamic clusters (Pinch *et al.*, 2003; Bathelt *et al.*, 2004).

UWM follows a typical hub (UWM) and spoke (start-ups) morphology. The most important exceptions to this are the firms that are genealogically connected to Promega, which is discussed further below.

Founded in 1848, UWM is a public university that was designated as a land grant institution in 1864. It was in the early 1900s that university president Charles Van Hise verbalised the concept of the "Wisconsin Idea". In keeping with the US land grant university tradition, he, and other members of the school's faculty and administration, believed that university knowledge should spread to the borders of the state or that, "the boundaries of the university are the boundaries of the state" (Sobocinski, 1999: 9). This established the continuing mission of "service to the state" (UWM Board of Regents 2003). For the last decade, UWM has been among the Top Five universities in the US in terms of R&D expenditures (see Table 7.5).

In 2004, the last year for which US National Science Foundation (2004) data is available, UWM (and the other University of Wisconsin campuses, which are statistically insignificant) expended in excess of USD 763 million. In 2004, the life sciences received USD 474 million while math, computer sciences, and engineering (MCE) received USD 113 million (NSF, 2004: 78) (see Table 7.6). The disciplinary expenditures show the strength of UWM in the life sciences. MCE at UWM is respectable, but it is not as strong as at some other Midwestern universities such as the University of Illinois. Not surprisingly, this strength in the biological sciences is expressed in the start-ups.

Table 7.5. **R&D expenditures by year (USD '000)**

Rank	Institution	1997	1998	1999	2000	2001	2002	2003	2004
	All institutions	24 370 716	25 856 006	27 530 968	30 068 664	32 793 818	36 367 358	40 056 637	42 945 081
1	Johns Hopkins University	829 241	853 620	874 518	901 156	999 246	1 140 235	1 244 132	1 375 014
2	University of California, Los Angeles	398 865	447 367	477 620	530 826	693 801	787 598	849 357	772 569
3	University of Michigan, all campuses	483 485	496 761	508 619	551 556	600 523	673 724	780 054	769 126
4	University of Wisconsin, Madison	419 810	443 695	499 688	554 361	604 143	662 101	717 044	763 875
5	University of California, San Francisco	343 384	379 970	417 095	443 013	524 975	596 965	671 443	728 321
6	University of Washington	409 959	438 191	482 659	529 342	589 626	627 273	684 814	713 976
7	University of California, San Diego	376 655	418 790	461 632	518 559	556 533	585 008	646 508	708 690
8	Stanford University	395 310	410 309	426 549	457 822	482 906	538 474	603 227	671 046
9	Pennsylvania State University, all campuses	339 955	362 643	379 402	427 575	458 066	492 739	533 427	600 139
10	University of Pennsylvania	296 141	333 477	383 569	430 389	469 852	522 269	564 635	596 756

Source: NSF 2006, Table 27.

Wisconsin Alumni Research Foundation

The Wisconsin Alumni Research Foundation (WARF) plays a critical role in the UWM ecosystem as an intermediary in the commercialisation of university research. It was established in 1925 as a non-profit patent organisation funded initially by UW alumni and managed by a Board of Trustees composed of alumni, but it has always been independent of the university. As such, and as a result of a historical accident relating to the patenting of ultraviolet irradiation of food to increase Vitamin D, it is a unique institution unlike that of any other US university. The independence allows it to operate in an entirely business-like fashion and it is not involved in university politics or managed by the academic administrators. WARF's primary purpose was and is to manage patents based on UWM research. Since 1928, WARF has provided more than USD 750 million to the University to support further research. As such, WARF has been a major force in the development and growth of the university as a research institution and in the entrepreneurial environment in Madison (WARF, 2007).

CLUSTERS, INNOVATION AND ENTREPRENEURSHIP – ISBN 978-92-64-04442-5 – © OECD 2009

Table 7.6. **R&D expenditures by discipline, University of Wisconsin, Madison, 2004 (USD '000)**

National rank	Field	R&D amount
29	Agricultural Sciences	43 238
6	Biological Sciences	155 682
17	Chemistry	17 115
23	Computer Sciences	13 457
14	Engineering	94 860
10	Environmental Sciences	54 127
7	Life Sciences	473 733
24	Mathematical Sciences	4 803
11	Medical Sciences	272 640
14	Physical Sciences	51 853
15	Physics	21 969
1	Psychology	29 329
4	Social Sciences	41 686

Source: NSF 2006, Tables 45-62.

WARF was successful nearly from its inception due to faculty assigned patents on the use of ultraviolet irradiation to increase Vitamin D in foods (in 1927), derivatives like coumadin and warfarin (a rat poison) (series of patents from 1941 through 1948), methods of preserving organs for transplantation (series of patents from 1967 through 1987), and magnetic nuclear resonance technology (WARF, 2007). WARF still receives between 60 and 70 per cent of its total income from Vitamin D technology (Gulbrandsen, 2003). Many of these lucrative patents were not commercialised through start-ups, but rather through licensing to large firms.

In 2005-06 alone WARF filed for 300 US patents on UWM technology and gave USD 65 million to UWM to support research. The process of patenting through WARF is described as:

> If WARF accepts the invention for patenting and licensing, the foundation provides an attorney to help the researcher with the patent application. The researcher also agrees to assign ownership of the invention to WARF. It is at this point WARF may contact companies considered good matches for the technology. WARF's policies call for 20 per cent of the gross licensing revenue from an invention to be returned to the inventor (or inventors). The remainder is shared with the UW Madison Graduate School, and the inventor's laboratory and department (WARF, 2007).

In addition to logistical support with the patent, WARF is able to help put scientists in touch with venture capital money (or other funding sources), offer loans and physical space for their company, and to provide advice and counselling in the early years. Deepak Divan, a UWM professor and founder of SoftSwitching Technologies describes WARF as "a gorilla standing beside you", who can enforce patents and act as an advocate (Ladwig, 2004).

The WARF support, and the connections, experience and funds it can provide help create an environment in which research and innovation are highly valued. This is especially important in the field of biotechnology, where the majority of advanced research is done at the university level, as opposed to in private or industry labs. Critical to the impact UWM has on the business landscape in Madison, and vital to the function of WARF, is the fact that professors retain the rights to the fruits of their research. This is unusual among research universities, as ownership tends to belong to the institution itself, not the individual. WARF has been able to capitalise on this fact.

Business school

Though it is debated as to what direct role business school programmes can have in assisting high-technology entrepreneurship, the UWM business school has become quite active in the field of entrepreneurship. (There are other programmes and departments within the UWM that have had roles in the development and evolution of the cluster.) A recent addition to the ecosystem is the Weinert Center for Entrepreneurship within the University of Wisconsin, Madison School of Business, which interacts with the high-tech firms in a novel way. Through the Weinert Applied Ventures in Entrepreneurship (WAVE) programme founded in 1998, 12 MBA students are selected each year to work with a new local firm. The students get experience while creating comprehensive strategic, operating and financing plans for the firm (Weinert Center, 2007). In return, the firm may benefit from the student's knowledge, a set of skills very different from those of the professors who are developing the technology. This programme is unique, and helps to create an environment where entrepreneurship is encouraged and enabled. The WAVE programme allows professors to market the products of their research while remaining an active member of the University's faculty by limiting their involvement in day-to-day business operations.

Though there have been no evaluations as to its impact, the Small Business Development Center, founded in 1979 and also a part of the University, offers courses in fundamental business areas. The SBDC's stated mission is, "[t]o enhance the success of small business owners and managers in our three county service area of Dane, Sauk and Columbia counties and encourage growth in our economy. We strive to achieve this mission by providing practical, customer-focused management education, training, counselling and networking" (SBDC, 2007). Class offerings range from what you need to know before starting a business to leadership skills and how to protect your firm in the case of a natural disaster. The organisation also has business counsellors available at no charge who can assist with specific issues.

Office of University-Industry Relations and Office of Corporate Relations

Established in 1963 as the University-Industry Research Program and renamed in 1994, the Office of University-Industry Relations was the campus link to small and large business, as well as the office that managed campus invention disclosures. Its mission was "to establish the most productive relationship possible between the private sector and the University of Wisconsin, Madison...". Some of the methods used to achieve that goal included campus tours of faculty research, the introduction of industry to UWM research consortia, and facilitation of partnerships through federal programmes such as Small Business Innovation and Research grants (SBIR). In 2005, the state of Wisconsin received USD 35 million in these federal grants for commercialising research (Wisconsin Department of Commerce 2005). As Table 7.7 indicates, the overwhelming bulk of this money went to Dane County where Madison is located.

Table 7.7. **State of Wisconsin SBIR grants by county in 2005**

Geographical distribution of awards	# of companies	USD amount	% of total
Dane County Alone	37	36 508 922	86.8
Dane, Sauk, Iowa Counties	39	36 649 227	87.1
Southeastern Wisconsin	10	3 414 004	8.1
Northwestern Wisconsin (Eau Claire)	1	456 965	1.1
Northeastern Wisconsin (Appleton)	1	69 345	0.2
Central/Northcentral (WI Rapids, Chili)	2	1 471 431	3.5
TOTALS	53	42 060 972	

Source: www.commerce.state.wi.us/NEWS/releases/2005/157.html (Accessed 30 January 2007).

Following a recommendation of the Chancellor's 2002 Task Force on University-Business Relations, UIR was phased out and replaced by The Office of Corporate Relations (OCR) in July 2003. The task force determined that the operating model of UIR was unable to keep pace with the rapidly changing needs of the business community in Madison and did not provide a visible, single point of contact for Wisconsin business. The new office, now located at the University's Research Park, is central point that provides clear access to a variety of programmes.

According to Allen Dines, Assistant Director at OCR, the role of the office is as "broker, connector, consultant and cheerleader" (personal interview, 2007). The OCR does not oversee programmes related to business start-up and industry support, rather it behaves as an interpreter of and liaison between these programmes and current businesses. It also works to connect new, start-up companies with the outside world. "We interpret the company's needs and refer them to University Programs *(ibid)*". In addition, the OCR serves as a public relations manager; Dines sees public relations as one of his, and his colleagues, major tasks as they act to "promote the idea that the University is friendly and open to business". It should be noted that, while the OCR is located in Madison and has a special interest in the city, their support of business reaches to the rest of Wisconsin as well as across the United States.

Research parks/small business incubators

Madison has experienced a proliferation of small business incubators. The principle behind business incubators is that new firms may be unable to mobilise the resources necessary for success.[8] The University supports one such incubator, but there are many others. The UWM-sponsored incubator, University Research Park (URP), is a non-profit entity established in 1984 that develops land to lease to start-up companies (Sobocinski, 1999: 306). The profits from this development are donated to UWM. There are currently 110 tenants, who have access to core services on such as accountants, lawyers and venture capitalists (Potera, 2004; URP, 2007).

The URP also houses the Madison Gas and Electric Innovation Center financed by the local utility and opened in 1990. The Innovation Center provides office and laboratory space for small companies, as well as shared support services and equipment. For companies beyond the start-up stage, but not ready or unable to move, the University Science Center at URP provides flexible space with room to expand. There is also land available at the Park where companies can build their own facilities. In early 2007, the URP plans to complete a second phase of development, which will add 270 acres and 53 buildings to the site, allowing an increase in the number of tenants to more than 200.

A second incubator, not associated with the University, is the TEC Incubator Center. It is geared toward technology firms and provides conference, classroom, and computer lab space as well as high-speed internet and phone service (TEC, 2007). The region is also home to the Fitchburg Technology Campus with a focus on nanotechnology research. Since the mid 1990s, Madison has experienced a proliferation of firm incubators not limited to the ones mentioned here.

Role of SMEs

The Madison cluster is a university-centric cluster. Whereas, biotechnology clusters like North Carolina will have large multinational corporations at their core, there are no such giants in Madison. Invitrogen, a biomedical input firm based in Carlsbad, California, employing 4 500 worldwide does have an operation

CLUSTERS, INNOVATION AND ENTREPRENEURSHIP – ISBN 978-92-64-04442-5 – © OECD 2009

in Madison. When understood from this perspective, entrepreneurial firms and SMEs drawing upon the university's research are the cluster. With the university, the SMEs are central to the cluster and new entrepreneurial firms are not unusual.

The success of small and medium sized businesses has reached critical mass and entrepreneurship has become an accepted path. Jane Homan, a UWM professor and co-founder of local biotech firm Gala Designs established in 1996 reflected upon the earlier ethos in Madison, "we were considered the mavericks of the University ... Now we're pretty normal" (Fikes, 2000). Absent these start-ups, it is likely that the Madison economy would be a typical public sector economy served by a small services-based private sector. Moreover, the state economy would have no growth regions or industries.

Barriers to cluster development

Despite the success of the region in creating start-ups, there are difficulties. The lack of airline connections has already been mentioned. A more important obstacle is a shortage of seasoned executives *in situ* that can be recruited to provide the business experience to the university-related innovators. The success of any start-up is predicated as much upon the ability to attract top-notch business talent as having first-rate scientists. An important recruiting ground for these seasoned executives are established firms within the industry. In biotechnology that would be the large pharmaceutical firms and in the IT sector it would be firms like IBM, Intel, and Microsoft; none of which have significant facilities in Madison or even Wisconsin. There can be little doubt that this is a handicap to Madison firms. Pam Christenson, acting administrator in the Business Development Division of Wisconsin's Department of Commerce believes workforce development is one of the biggest barriers facing the cluster to this day (Christenson, 2007). The University and the business school have responded to this shortage by creating various programmes such as entrepreneurship and technology management specialties and a master's of science in biotechnology, but ultimately these cannot overcome the lack of large firms with seasoned management in the region.

A lack of venture capital funding is also cited as an obstacle to the growth of the cluster. However, Allen Dines of the OCR has a slightly different perspective. For him, the obstacle has been the ability to create "fundable deals that are attractive to investors" (Dines, 2007). Many Madison start-ups have not sufficiently clearly identified their markets as they make the transition from technology to product. Were this identification more clear, then he believes the search for funding would be more successful. He also notes that most of the Madison start-ups are not attractive to VC funders, because the types of products and services they intend to produce will not serve sufficiently large markets. Conversely, the small specialised (but very lucrative) markets do not require such large expenditures in the establishment of the firm.

The barriers to the further development of the cluster are difficult to predict. Obviously, one barrier is the number of highly-educated university faculty and graduates that have developed technology worthy of commercialisation. Further, if they have developed technology worthy of commercialisation, do they wish to be involved in commercialisation. In terms of technologies and markets, it seems likely that there will be a continuing flow of opportunities to commercialise various research technologies, materials, and services. For example, a UWM professor through WARF holds key patents on stem cell lines, patents that may create an ample income source, but which have also created much controversy due to the severe licensing restrictions WARF has imposed on industrial and academic researchers.[9]

Role of policy

Together with the University and WARF, public policy plays a role in the development and maintenance of the high-tech cluster in Madison, Wisconsin. What follows is a discussion of some of these policies.

Federal policies

The US is quite different from other nations in that the Federal government has had very few region-specific policies with the possible exception of the major support for defence firms and bases on the West Coast during World War Two. The US government's most significant policy for UWM has been enormous and unremitting funding of university research particularly the life sciences; some small portion of the research results are commercialisable. In the case of UWM, the most important research funding has been in the biomedical and, to a lesser degree, agricultural fields. UWM has not been one of the elite US computer science and electrical engineering schools so DARPA funding was not as significant for the local start-up economy.

Bayh-Dole is credited by some such as Howard Bremer, emeritus patent counsel at WARF, as being important to the development of a commercialisation ethic at UWM, though it should be noted that many of WARF's greatest successes came long before Bayh-Dole was signed. UWM was the first school to sign an Institutional Patent Agreement (IPA) with the then US Department of Health, Education, and Welfare that granted UWM the right to license inventions to firms without having to clear it with the federal government (WARF, 2007). Use of IPAs quickly spread to other major US research universities

 CLUSTERS, INNOVATION AND ENTREPRENEURSHIP – ISBN 978-92-64-04442-5 –

(Mowery *et al.*, 2004), and they were the precursor to the Bayh-Dole Act of 1980 that was a blanket transfer of the rights to federally-funded university inventions to universities.

The passage of Bayh-Dole and the success of firms, such as Genentech, Amgen, Hybritech, and many others, in commercialising the new biotechnologies that had been developed in university laboratories piqued the interest of university administrators searching for new sources of income. The result was a rush of universities establishing offices for technology licensing. At UWM, WARF was already established for commercialising university faculty inventions and, unsurprisingly, it moved into this field. One change in the environment was that increasingly professors wanted to be involved in the commercialisation of their inventions through establishing a firm. And, frequently, the firm was located close to the university. It is this dynamic that led to the emergence of a local cluster of technology firms in Madison.

The SBIR programme provides about USD 35 million to the Dane County economy, and has been significant in the growth of some firms. Since SBIR funds substitute for early stage venture capital, and the Madison area has only minimal amounts of venture capital, the SBIR programme probably is of some importance to entrepreneurship in the Madison. Despite the debates about the overall efficacy of the SBIR programme, it is likely to have been important to certain firms in the region.

State and local policies

While the University and WARF are central to the creation of the cluster, local and state government have provided the infrastructure and support necessary to maintain its growth. One way this is pursued is through groups such as the Wisconsin Technology Council and the Wisconsin Biotechnology and Medical Devices Association, who lobby the federal and state governments in the interest of local firms. These organisations also help with regulatory issues. The state responds well to such lobbying efforts. In 2004, Governor Doyle announced that Wisconsin would invest up to USD 750 million in biomedical research over the next few years. One such programme included in that pledge is the planned Wisconsin Institutes for Discovery. Construction began in 2008 for the public-private partnership that includes a large private donation well as matching gifts from WARF and the state of Wisconsin. The Institutes will be housed on the UWM campus and have as their goal the fostering of interdisciplinary collaboration and innovation.

Outside of Madison, many of the cities in Wisconsin have faced economic difficulties in recent years. State wide, the Grow Wisconsin Initiative looks to reverse this trend by creating an environment that encourages business development. The plan laid out by the governor in September 2003 focuses on four key areas: 1. the creation of competitive business climate; 2. investment in the people of Wisconsin; 3. investment in Wisconsin businesses, and; 4. the reform of regulations and increased government responsiveness.[10] It is too early to gauge the results of this initiative that plans to invest over one billion dollars to achieve its goals, including USD 300 million in seed and venture capital funds and another USD 10 million for a free training fund for companies looking to invest in new high-tech jobs. Regardless of this, the support of the government is impressive.

Unlike many other regions with high rates of start-up creation, there are limited venture capital funds available in Madison. In 1998, there were 40 states that recorded venture capital activity. Only eight of these had lower levels of activity than Wisconsin. Rates of VC funding have increased since that time, but they remain well behind those typical of regions with high levels of technology-based entrepreneurship. Supported by local and state policies, angel financing has been able to fill this void. The first angel investing group in the state of Wisconsin was founded in 2000. Today, there are 15 networks (Table 7.8).

Table 7.8. **Angel investment groups in Wisconsin**

Firm name	Location(s) in Wisconsin
Badger AgVest, LLC	Wausau
Badger Alumni Capital Network (BACN)	
Central Wisconsin Business Angels	Wisconsin Rapids
Chippewa Valley Angel Investors Network	Eau Claire
IQ Corridor Angel Network	Pewaukee
Marshfield Investment Partners, LLC	Wausau
New Capital Fund, LP	Appleton
Origin Investment Group	LaCrosse
Pennies from Heaven	Racine Country, Kenosha County
Phenomenelle Angels Fund I, LP	Madison
Silicon Pastures	Milwaukee
St. Croix Valley Angel Network	River Falls
The Golden Angels Network	Milwaukee
Wisconsin Investment Partners, LLC	Madison
Women Angels	Milwaukee

Source: NorthStar Economics, Inc. 2007.

Indicative of the willingness of state and local government to support the biotech cluster and angel financing is the recent adoption of the Wisconsin Angel Tax Credit (Act 255). Beginning in January 2005, the state of Wisconsin made USD 3 million available annually in angel tax credits. The full amount

CLUSTERS, INNOVATION AND ENTREPRENEURSHIP – ISBN 978-92-64-04442-5 – © OECD 2009

was expended in 2005 and just over USD 900 000 remain from 2006 (Commerce Department, 2007). While analysis on the effects of the tax credit is still incomplete, initial findings are optimistic. According to a report filed by NorthStar Economics, Inc., investing by angel groups in 2005 increased by 65 per cent over the previous year. Of course, the ultimate test is whether high-quality firms were created, thereby justifying the state investment.

On a local level, ordinance amendments have been passed to alter zoning restrictions for companies looking to set up biotechnology research facilities. In addition the city provides tax incremental financing for businesses locating in specific districts as well as revolving loans. Together with three other counties in Wisconsin, Dane county is involved in a programme that offers tax credits for new and expanding high-technology businesses (Office of Business Resources, 2007).

While the policies and programmes mentioned above have played an important role in the cluster, it is important to realise that these influences can only do so much. Pam Christenson from the Wisconsin Department of Commerce points out that "we see clusters as a private sector driven initiative, not a programme the public sector can impose on an industry" (Christenson, 2007). Her office produced a white paper in 2003 entitled *Fostering Cluster Development in Wisconsin*. The publication reads, "the private sector must lead successful clusters. Business and other key stakeholders should examine the changes and improvements that need to occur *within* the cluster and not focus solely on what government should do *for* the cluster" (Wisc. Dept. of Commerce 2003: 2).

Future policy challenges

For a state like Wisconsin, which has felt the full brunt of deindustrialisation, research-based entrepreneurship has been seen as an important boon to the state. The success has been so great that recently the state has been asking UWM and WARF to assist other cities in Wisconsin to develop technology clusters. From the political perspective, this makes perfect sense. However, there is a possibility that the institutions at UWM and in Madison will lose their focus on growing the cluster in Madison, thereby inhibiting its growth, while it is unlikely that the other regions have the technological bases to establish viable clusters and, perhaps, even viable firms. Thus its very success might lead local and state politicians to extend UWM's mission in directions that divert it from doing what it does best, which is grow its local cluster.

Tom Still of the Wisconsin Technology Council is quoted in a recent newspaper article on the future challenges facing high-technology start-up business in Wisconsin."The bottom line is Wisconsin is seeing more activity –

that doesn't mean we're where we need to be; we're well behind the curve, but at least we're laying a strong foundation for stronger investment activity years ahead".

The Madison biotechnology cluster is significant. However, government and industry leaders believe that growth, especially outside of the supply and services fields, will depend on increased venture investment – both in quantity and value of deals. Whether this belief is justified or not is beyond this chapter's scope. Unfortunately, the prospects for more local venture capital do not look particularly promising because the UWM start-ups do not offer the promise of the enormous returns of a blockbuster drug. In addition, the region does not have many IT-related start-ups that provide extremely the rapid returns that ensure high internal rates of return for the venture capitalists.

Taxation is often mentioned in Wisconsin as an issue for the health of the cluster. A recent study for the Small Business Administration found that Wisconsin has above-average tax levels. While the Commerce Department notes that business taxes are lower in Wisconsin than those in 35 other states, an anti-tax group, the Wisconsin Taxpayers Alliance, reports that state and federal taxes claimed 33.4 per cent of personal income in the state on 2006 (Still, 2007). Some believe that if the state wants to encourage the continuing development of the high-tech cluster, these supposedly high rates of state taxation should be lowered. It should be noted that the most successful state in the US in terms of high-technology entrepreneurship, California, has a roughly similar tax burden. Also, given that Wisconsin's start-ups are so university-linked, it is unlikely that these entrepreneurs would relocate to a state with very low tax burdens such as Mississippi. Tax cuts that weakened UWM would almost by definition in such a public university-centric cluster have a negative effect on the cluster. Those arguing that tax cuts would strengthen the development of the Madison cluster likely are more interested in tax cuts than in the furtherance of cluster growth.

The greatest policy challenges for the cluster centre upon any changes in the levels of Federal funding for research. The current federal deficit spending, particularly on the invasion of Iraq and the military, could lead to a situation within which federal research funding decreased. Because of the centrality of UWM research to overall cluster health, such an event would almost certainly weaken the cluster. The impact would be magnified if biological research were particularly singled out for cutbacks. The dependence of the cluster on federal funding of biological research, both for continuing growth both in existing firms, many of which produce biologicals for research, and the flow of new start-ups cannot be underestimated.

CLUSTERS, INNOVATION AND ENTREPRENEURSHIP – ISBN 978-92-64-04442-5 – © OECD 2009

Lessons for other clusters

It is hazardous to draw too strong conclusions from a single case study. However, Madison's experience does suggest the following observations:

1. High-technology entrepreneurship has been important for the growth of the Madison economy, and has made it the fastest growing part of the state of Wisconsin.
2. The development of the Madison cluster has been underway for approximately 25 years, and it has developed a regional recipe that does not require large amounts of venture capital. A number of firms have received venture capital. The exact chronology of this co-evolution between technology fields and financing strategies is not clear, but in 2007 it is now conventional wisdom among entrepreneurs.
3. UWM has a unique institutional relationship with the fully independent WARF, which is responsible for all licensing and technology transfer. This contrasts with dominant model in US and increasingly universities around the world where the technology transfer organisation is a part of the university. The success of university entrepreneurship and WARF in returning funds to the university suggests that the dominant model may not be best for all universities, and that experimentation with other models might be valuable.
4. For university-based clusters, national-level decisions on research funding may be as important as any local or state decisions particularly in smaller less wealthy states such as Wisconsin.
5. As Klepper and Sleeper (2005) and others have discovered, one firm such as Promega in Madison or Fairchild in Silicon Valley may be very important as a source of still further entrepreneurial firms.
6. UWM's research excellence particularly in biology has translated into local economic development.
7. The state government has invested in a large variety of initiatives to encourage the growth of the cluster. However, there have been few evaluations of the efficacy of this funding. The most successful UWM firms such as Promega received little direct assistance from these new state programmes as WARF has historically been the locus of university-based spin-off activity.
8. The living conditions retain Madison entrepreneurs. Unfortunately, Madison has been less successful in attracting either top-flight executives to the region or entrepreneurs that established their firms elsewhere. Thus it falls in between regions such as Pittsburgh with universities like Carnegie Mellon that often lose their entrepreneurs to other regions such as Silicon Valley or Boston, and universities and regions such as Stanford/UC Berkeley and regions such as Silicon Valley that attract entrepreneurs from around the world.

As was the case in Silicon Valley and many other high-technology clusters, Madison suggests that it was not far-sighted policy makers that were responsible for the creation of the cluster, but rather pioneering entrepreneurs. Moreover, the Madison entrepreneurs did not compete in human therapeutics, which required large tranches of venture capital and were extremely risky, but rather found niches that could be bootstrapped into profitable small businesses. In the overall scheme of things, the Madison cluster is relatively small, however for the medium sized city of Madison it is an important component of the city's overall success.

Ultimately, the inception of every cluster is based upon entrepreneurship and, rarely, is this an outcome of government policy. Usually, policy follows the emergence of a cluster, and hopefully it does not retard the cluster's growth. One of the greatest mistakes by practitioners, policy makers, and academics is to examine an established cluster and the extant policies at that time and assume that the policies were responsible for the birth and growth of the cluster. It is only through longitudinal analysis that we can surely identify the reasons for the birth of the cluster and evaluate the reasons for its growth.

Notes

1. This detailed information was provided by Allen Dines, Assistant Director, Office of Corporate Relations, UWM.
2. On entrepreneurial support networks, see Kenney and Patton 2005.
3. On the dynamics of venture capital investment see Florida and Kenney (1988) or Gompers and Lerner (1999).
4. *www.wisgov.state.wi.us/docview.asp?docid=707&locid=%2019.*
5. Kenney and Patton 2005, Powell *et al.* 2002.
6. Our research has strict definitions of high-technology. For example, while some lists of technology firms in Madison include personal computer assembly shops or routine blood testing firms, we exclude these. For this reason our list is shorter than the one produced by Madison Gas and Electric.
7. This may be politically correct, but may also be a waste of UWM-affiliated persons' time even if the resources are provided by the state.
8. The belief by the supporters of incubators is that they can shepherd small firms over the initial difficulties typical of firms. Another school of thought, frequently held by successful entrepreneurs, argues that locating firms to an incubator merely prolongs the lives of unworthy firms and might harm good firms by not exposing them to market rigors immediately.
9. Initially, UWM attempted to charge royalties to any researchers wishing to do stem cell research. This policy created enormous criticism and in late January 2007 it modified the policy to allow researchers to use the materials and techniques royalty free.
10. *www.wisgov.state.wi.us/docview.asp?docid=707andlocid=%2019.*

References

Bathelt, H., A. Malmberg and P. Maskell (2004), "Clusters and Knowledge: Local Buzz, Global Pipelines and the Process of Knowledge Creation", *Progress in Human Geography*, Vol. 28, No. 1, pp. 31-56.

Christenson, Pam (Acting Administrator, Department of Business Development, Wisconsin Department of Commerce) (2007), "Personal Email Communication with Amanda Nelson", 30 January.

Dines, Allen (Assistant Director, Office of Corporate Relations, University of Wisconsin-Madison). (2007), "Personal Interview by Amanda Nelson", 24 January.

Feldman, M.P., J. Francis and J. Bercovitz (2005), "Creating a Cluster While Building a Firm: Entrepreneurs and the Formation of Industrial Clusters", *Regional Studies*, Vol. 39, No. 1, pp. 129-141.

Fikes, B. (2000), *Location, Location, Location? Biotech Moves out of Traditional Hubs,www.doubletwist.com*, accessed 24 July.

Florida, R. (2002), *The Rise of the Creative Class*, Perseus Book Group, New York.

Gompers, Paul and Josh Lerner (1999), *The Venture Capital Cycle*, MIT Press, Cambridge.

Grow Wisconsin Initiative, *www.wisgov.state.wi.us/docview.asp?docid= 707&locid =%2019*.

Gulbrandsen, Carl (managing director, WARF) (2003), "WTN Interview: Carl Gulbrandsen WARF – A Unique and Successful Technology Transfer Organization", *http://wistechnology.com/article.php?id=416*, accessed 14 January 2007.

Kauten, R. (2006), PanVera Founder, "Personal Email Communication with Amanda Nelson", 3 March.

Kenney, M. (1986), *Biotechnology: The University-Industrial Complex*, Yale University Press, New Haven.

Kenney, M. and D. Patton (2005), "Entrepreneurial Geographies: Support Networks in Three High-Tech Industries", *Economic Geography*, Vol. 81, No. 2, pp. 201-228.

Klepper, S. and S. D. Sleeper (2005), "Entry by Spinoffs", *Management Science*, Vol. 51, No. 8, pp. 1291-1306.

Ladwig, Jill (2004), "Climbing Down the Ivory Tower", *Madison Magazine*, July, *www.madisonmagazine.com*.

Madison Gas and Electric Company and the City of Madison (2004), "Greater Madison Wisconsin Area Directory of High-Tech Companies".

Madison Gas and Electric Company and the City of Madison (2006), "Greater Madison Wisconsin Area Directory of High-Tech Companies".

Madison Gas and Electric Company (n.d.), brochure entitled "What Makes our Area so Appealing to over 500 High-Tech Businesses?".

Madison Gas and Electric Company (2007), *www.mge.com*.

Mowery, D. C., R. R. Nelson, B. N. Sampat and A. A. Ziedonis (2004), *Ivory Tower and Industrial Innovation: University-Industry Technology Transfer before and after the Bayh-Dole Act*, Stanford University Press, Stanford.

National Science Foundation (2006), *Academic Research and Development Expenditures: Fiscal Year 2004*, National Science Foundation, Washington, DC, *www.nsf.gov/statistics/nsf06323/pdf/nsf06323.pdf*.

Office of Business Resources (2007), City of Madison, Wisconsin, *www.businessmadison.com*, accessed 29 January 2007.

Office of Corporate Relations (2006), *The Office of Corporate Relations Annual Report for 2005-2006.*

Pinch, S., N. Henry, M. Jenkins and S. Tallman (2003), "From 'Industrial Districts' to 'Knowledge Clusters': A Model of Knowledge Dissemination and Competitive Advantage in Industrial Agglomerations", *Journal of Economic Geography,* Vol. 3, pp. 373-388.

Potera, Carol (2004), "America's Third Coast Nurtures Biotechnology", *Wisconsin Technology Network, www.wisctechnology.com,* accessed 8 March 2007.

Rosen, Michael (2006), "Midwest Plays Small Role in VC Investment, Angel Groups to the Rescue", *Wisconsin Technology Network,www.wisctechnology.com,* accessed 20 February 2007.

Small Business Development Center, *www.uwsbdc.org*, accessed 20 January 2007.

Sobocinski, P. (1999), *Creating High-Tech Business Growth in Wisconsin: UW Madison, Technology Transfer and Entrepreneurship*, University-Industry Relations.

Still, T. (2007), "High Taxes and Low Rates of Entrepreneurship – Is there a Relationship?" Wisconsin Technology Network, *www.wistechnology.com*, accessed 8 January 2007.

Tatge, M. (2004), "Miracle in the Midwest", *Forbes,* Vol. 173, No. 11, pp. 120-128.

TEC Incubator, *www.tecenterinc.org*, accessed 9 February 2006.

University Research Park, (2007), *www.universityresearchpark.org*, accessed 27 January 2007.

UWM Board of Regents (2003), *Report of the Chancellor's Task Force on University-Business Relations, www.chancellor.wisc.edu*, accessed 21 April 2003.

WARF (Wisconsin Alumni Research Foundation) (2007), *www.warf.ws*, accessed 14 January 2007.

Weinert Center, WAVE Program, *www.bus.wisc.edu/weinertcenter/wave/.*

Willis, Randall (2004), "America's Biotech Heartland", *Modern Drug Discovery,* American Chemical Society, pp 41-46.

Wisconsin Department of Commerce (2003), *Fostering Cluster Development in Wisconsin,* Agency White Paper.

Wisconsin Department of Commerce (2005), "Governor Doyle Announces Recipients of SBIR/STTR Awards", *www.commerce.state.wi.us/NEWS/releases/2005/157.html,* accessed 30 January 2007.

Wisconsin Department of Commerce (2007), *www.commerce.state.wi.us,* accessed 29 January 2007.

ISBN 978-92-64-04442-5
Clusters, Innovation and Entrepreneurship

Chapter 8

The ICT Cluster of Waterloo, Canada

by
David A Wolfe, Ph.D. Program on Globalisation
and Regional Innovation Systems,
Centre for International Studies, University of Toronto

This chapter illustrates the approach and initiatives of the entrepreneurial university in Waterloo in building the region's information and communications technology (ICT) cluster. The creation of spin-offs seems to be encouraged by a series of policies implemented in the university, among students and academia. This case study is also a good example of collaboration between university and industry to meet the demands of the cluster labour market concerning skills. The adaptation of the university to the needs of industry, and the tight co-operation between industry and university in training and developing specific skills in students, have had a positive impact on the development of the cluster. This chapter also shows how flexible intellectual property regulations at the university play an important role in stimulating innovation, collaboration and business creation.

Introduction

The search for effective local economic development strategies has assumed greater significance over the past decade. Concern with the factors that sustain innovation in local and regional economies has led to a growing fascination on the part of policy makers with industrial clusters due to their perceived impact on competitiveness and innovative performance. Closer examination of the origins and development path of specific clusters provides some guidance for cluster policy. Also important is the relationship between large and smaller enterprises in clusters. Much of the cluster literature documents the role of lead anchor firms in grounding the cluster in a specific geographic location. At the same time, a key success factor in cluster development is the ability to spin-off and grow newer small and medium-sized enterprises (SMEs). Much of the cluster literature views large and smaller scale firms as integrally linked in terms of buyer-supplier relationships, with the smaller scale firms forming a critical part of the supply architecture of the regional economy. Yet evidence from more recent studies indicates that buyer and customer relationships with external partners can be equally important for cluster firms as their internal relationships. A number of these studies point to the role of niche clusters within the global economy, as individual clusters in photonics, for instance, have come to specialise in particular technology and sectoral niches. Another key issue involves the relationship between different levels or scales of governance in supporting cluster policy. While clusters are clearly defined as a local or regional phenomenon, there is growing recognition that vibrant and dynamic clusters draw effectively on a range of policy instruments from senior levels of government to sustain their development.

The case of the information and communication technology (ICT) cluster in Waterloo, Ontario, is instructive for developing a better appreciation of the implications of these factors for the process of cluster formation. This chapter draws upon a recent case study of the Waterloo, Ontario, ICT cluster – conducted as part of a larger Canadian study of industrial clusters – to explore these issues. The following discussion examines the origins of this vibrant regional economy and explores how a deeply rooted regional culture, historical patterns of trade and knowledge flows, and locally created institutions each contributed to the emergence of the region as a dynamic centre of high-tech activity. Along the way, it provides some valuable insights for policy makers interested in emulating the region's cluster-based success.

 CLUSTERS, INNOVATION AND ENTREPRENEURSHIP – ISBN 978-92-64-04442-5 – © OECD 2009

Nature and evolution of the cluster

The ICT cluster in the Kitchener-Waterloo-Cambridge (Waterloo) region, located an hour west of Toronto, is one of the most dynamic sources of high-tech activity in Canada. Although the present contours of the cluster can be traced back to the formation of the first software and computing firms in the 1970s, the Waterloo region has long been an important location for manufacturing in the Southern Ontario industrial landscape. Kitchener-Waterloo has been the home to major national and international corporations for more than a century, from Dominion Electrohome Ltd. to present day success, Research in Motion Inc. – manufacturer of the iconic "Blackberry". The region has had a pioneering presence in some of the major technological advances in North America, including automobiles, radio, processed foods, financial services, biotechnology and computing. Today, this history of technological leadership continues in fields such as internet-enabled wireless communications, software, aerospace, engineering, ecommerce, robotics, and laser technology.

The emergence and contemporary dynamism of the Waterloo ICT cluster owes its success to critical decisions taken by industrial leaders in the local economy in the years following World War II. The period between the wars saw the growth of complex engineering, metalworking, food and automotive-related industries in the region on the foundation of the traditional manufacturing base. One of the key local institutions to emerge in the interwar period was Waterloo Lutheran College, established in Kitchener in 1924. Although the college did not contribute directly to the high-technology development in that era, its offspring, the Associate Faculties, was the precursor to the University of Waterloo. The University of Waterloo is one of several colleges and universities in the region with strong ties to local industry. However, more than any other research institution in the region, it has exerted a singular impact on the regional economy. The period of post-war reconstruction in Canada and the growing recovery of the leading industrial powers brought home some important lessons for government and industry in Canada. In a world where national survival was predicated on technological capabilities, Canada was found woefully lacking by industrialists and government alike. Local leaders and institutions in Kitchener-Waterloo played a key role in translating those lessons into practical measures. The University of Waterloo, founded in 1957, emerged in response to the growing demand for more sophisticated and technical educational institutions.

Certain prominent members of the local industrial community in Kitchener-Waterloo played a strategic role in Waterloo Lutheran College through their membership on the Board of Governors and recognised the growing demand for trained technical personnel and the implications of this demand for the employment requirements of the local economy. Recognising the need for more technical education in the regional economy, Ira Needles (president of B.F. Goodrich and chairman on the Board of Governors for the newly created Associated Faculties of Waterloo Lutheran College) proposed a unique solution in the form of *The Waterloo Plan*. This plan called for a new type of education to be offered on a co-operative basis with industry. In sharing the burden of technical training with industry, the university would be able to support twice the number of students (as one class rotated out to co-op placements in industry while another took its place in the classroom), provide a greater depth of education – both theoretical and practical – and build a closer relationship with industry in order to anticipate employment needs, secure additional funding and ensure that classroom education remained on the cutting edge. This proposal became the basis for the University of Waterloo's highly successful co-operative education programme, widely regarded as the largest and the best university co-op programme in North American and a significant asset to the region.

Chance also played an important role in the early development of the ICT cluster – as it often does. The original plan incorporating the Associated Faculties assumed that it would remain affiliated with Waterloo Lutheran College, which would provide the liberal arts and social science components of the new university's curriculum. However, when the Associated Faculties acquired university status, the original college decided not to participate in the new institution. This serendipitous development resulted in the establishment of the new university with the overwhelming bulk of its faculty and course offerings in the sciences, math and engineering. Waterloo is one of the few universities in North America with a dedicated Faculty of Mathematics. In its formative period the university was mainly concerned with training a pool of local talent and transferring knowledge to the local economy through its graduates. It set out to provide the best possible science, math and engineering curriculum possible. The co-operative education programme, adopted in part out of financial necessity and in part out of the foresight of its founders, rotates students to industry and back to the classroom on a regular basis. This reflexive relationship allows the curriculum to keep up with the ever-changing technological frontiers of industry, while strong industry support for the programme has funded the acquisition of technology to enhance classroom learning. It was thus that Waterloo became one of the first universities in Canada to enable students to actively explore and make use of innovations in the relatively new academic field of study – computing.

A key development in the emergence of the cluster occurred with the installation of the first electronic computer. In the late 1950s, Wes Graham was recruited to the university from IBM to teach a statistics course. With his background in computing, Graham quickly became involved in a project to launch the new discipline of computer science at the University. The first computer arrived from IBM in 1960 – at a time when there were just over one hundred installed across the country – and it became the foundation of a computing centre that was continuously upgraded. By 1967, the University had an IBM 360/75, the largest computer in Canada. The first major ICT breakthrough at the University was an innovation in software – the WATFOR compiler. As soon as it obtained its first computer, the engineers who wanted to, and the mathematicians who could, started developing software. The only language available that allowed undergraduates to programme computers was FORTRAN, but it was too inefficient for practical use by large numbers of students and faculty. Faced with this limitation, students and faculty at the University invented the Waterloo FORTRAN compiler to speed up programming computations. This new technology, dubbed WATFOR, became the basis for one of the University's first spin-off companies and the first software company in Waterloo – WATCOM (1974) – now the parent company to several generations of subsequent spin-offs in the ICT cluster. The WATCOM spin-off established the basis for a new business model of the relationship between the company and the University. It allowed the founders of the company to retain ownership of their research and intellectual property and thus formed the basis for the University's current intellectual property policy. Furthermore, it provided an important example of the entrepreneurship which served as both a model and a stimulus for successive generations of university spin-off companies.

Geographically, Canada's Technology Triangle encompasses the four municipalities of Waterloo, Cambridge, Kitchener and Guelph. Overall, the region boasts 455 companies involved in the high-technology sector. The companies are spread across four sub sectors: information and communication technology, scientific and engineering services, advanced manufacturing, and the life sciences biotech and environmental sub sector. Of these, information and communications technology accounts for 62 per cent of the high-tech firms and employs 13 000 people or 45 per cent of the total in the high-tech sector (Communitech, 2005). Though there may be several firms involved in a particular market segment or technology niche within the region, they rarely compete directly with one another. This is a testament to the incredible diversity of high-tech activity in the region. The competitive advantage of firms is the uniqueness of their products. Since these products are so highly differentiated, most firms in the region compete globally on the basis of this technical excellence, rather than on cost.

Despite the relatively small size of the local community – the population of the Waterloo region was 438 515 in the 2001 census – the ICT cluster ranks among the top ten among census metropolitan areas (CMAs) in Canada and among the top thirty in North America on most indicators. The ICT manufacturing and service clusters are far from the largest in terms of number of employees, with 7 165 and 11 615 in the 2001 census. However, their combined employment ranks second behind the automotive parts cluster in the regional economy and the location quotients for each are well above 1.00 (Tables 8.1). Where the ICT sector makes its most significant contribution to the regional economy is in value-added. In a very conservative measure of ICT firms in the region, Canada's Technology Triangle reported that in 2000 ICT companies generated over CAD 8 billion in revenue. Furthermore, between 1993 and 1999 this sector's revenue increased 120 per cent, assets increased by 163 per cent and equity increased 420 per cent, indicating strong actual and potential growth (Canada's Technology Triangle, 2004). While export figures don't specifically target ICT-intensive industry, the research also indicates that most of the ICT firms in the Waterloo cluster produce almost exclusively for North American and global markets. The majority of measured exports from the region came from advanced manufacturing – including ICT-intensive – firms. In 2000, the region exported CAD 8.9 billion worth of products, 55 per cent of the region's GDP that year. Export activity in Waterloo is so significant that measured by the dollar value of exports per employee it ranks third in comparison to all US metropolitan areas (Canada's Technology Triangle, 2004).

Table 8.1. **ICT manufacturing and services**

ICT	Kitchener		Ontario		Canada	
	Manufacturing	Services	Manufacturing	Services	Manufacturing	Services
Number of establishments	138	400	3 357	13 348	7 813	28 420
Total labour force	7 165	11 615	134 375	266 285	242 950	628 885
Average establishment size	52	29	40	20	31	22
Location quotient – employment	2.00	1.25	1.44	1.10	1.00	1.00
Average annual income (CAD)	43 648	43 349	48 942	48 241	45 589	44 445
Compound annual growth, 1998-2005 (%)	2.9	8.7	1.2	6.2	1.4	4.9
% Full-time employment	97.4	85.9	95.9	89.2	95.3	89.2
% Part-time employment	2.8	13.8	4.1	10.8	4.7	10.8
% Self-employment	1.3	3.7	1.3	6.1	1.5	5.3
Average age of labour force	39.2	37.8	38.3	37.6	37.8	37.8
% foreign born	30.1	25.4	43.3	35.6	35.0	26.5

Source: Statistics Canada, Census of Canada, 2001.

The nature of the commercialisation process in the region, and in particular, the role of the University of Waterloo as a key institution in transferring new knowledge into the region has evolved considerably over the period. Whereas it played a more direct role as a knowledge generator in the 1970s and 1980s, the number of spin-offs and the results of a social network analysis (Xu, 2003) indicate that the degree of knowledge transfer through new firm formation has declined subsequently. Although the University remains central to the continuing development of the cluster, its primary contribution is no longer through the process of new firm formation. Relatively fewer firms have spun out directly since the late 1980s and the post-2000 slump in the demand for high-tech products and services resulted in a noticeable decrease in the availability of financing for start-ups and spin-offs. While the post-2000 downturn in the ICT sector has clearly had a negative impact on the regional economy, on the whole, it has much fared better than some of the other high-tech clusters in Ontario, such as the Ottawa-Gatineau region.

Success factors

The University of Waterloo continues to play three critical roles in the development of the region's ICT cluster. As a major research university, it is at the forefront of knowledge creation in a variety of fields. It also generates a key supply of talent that has contributed to the growth of a "thick" labour market in the local economy. Finally, through the process of knowledge creation and its strong support for entrepreneurship, the university has spun off several prominent firms in the area. While all three roles have had important effects on the shape of the cluster today, the one which attracts the most obvious attention to the local cluster is its role in spinning off high-tech firms. University or public research organisation spin-offs have long been a key goal of public policy makers and economic development officials. For one, they indicate the presence and creation of commercially viable research within a publicly funded institution and are therefore a mark of institutional success as well as a potentially positive return on public investment.

Several aspects characterise the nature of interaction between local high-tech firms and the University of Waterloo. First, while there are many formal relationships such as research contracts and funding of research chairs, much of the knowledge exchange is more informal than formal. Interviewees cite the University not only as an important source of tech transfer and specialised skills, but also as providing both international cachet to the region, and simple social/professional networks; "I contact my friends there if I have a problem". This informal approach underscores the "embeddedness" of the University in the local community, and many people emphasise the organic nature of the impact that the University has on the local community through the interaction effect between the various roles of the University – R&D transfer, skills provision, international cachet, and informal "knowledge networks".

There is, however, a distinct division between those firms that interact with the University and those that do not. For those with linkages to the University, there is a range in the depth and breadth of interaction. Larger firms tend to have more robust partnering relationships, often involving the funding of research chairs, long-term collaborative research projects, university faculty working within the firm, and full-time staff occupied with university and government interaction. Smaller firms, in contrast, tend to engage in short-term, problem-focused research projects. One of the primary reasons cited for not becoming more involved in university research is the amount of time required for commercialisation; while university research project horizons might be two to three years, firms often "need to work on things that need to be commercialised in 6-18 months". Others reported difficulty in accessing what was available, not feeling "in the loop", or had a perception that the research efforts at the University were focused on larger companies. Regardless of involvement with the University on an R&D level, however, almost every firm cited its critical importance as a provider of highly skilled and specialised talent.

The case of the Waterloo ICT cluster confirms that the presence of a robust local talent pool or "thick" labour market is indeed a central factor in the internal dynamics of the local cluster, and that the local college and universities have been the key actors in its development. A key explanation that firms provided for why they are located in Waterloo is that they have come to rely on such advantages as the local labour pool and the international cachet that the area carries in tech circles. They stress the interdependence of several key factors; the most often cited are the presence of the local universities, and the quality of the local talent pool. Most firms indicated that it was a distinct advantage to be located in Waterloo because it provided a ready supply of "smart and competitively priced" engineers and because the University of Waterloo is "one of the best universities in the world for computer engineering". In terms of the relative cost of building software in India, the US, Europe and Canada, Canada, and specifically Waterloo, is seen to be one of the best locations because of the quality and productivity of the local talent pool relative to its cost. The presence of large software and technology-intensive firms in the area, serves as a magnet and an anchor for the highly specialised labour pool. Firms stay in the area because they have invested in the local talent pool through in-house training which has generated tacit knowledge that is difficult to relocate. Some smaller firms indicated that it is "important to be close to [large anchor] companies that are leading edge" because they deepen the labour pool, and smaller firms can hire people that "used to work at RIM ... Descartes ... MKS, and others".

The University of Waterloo is considered to be the premier educational and research institution in the cluster, but it is only one of the post-secondary

 CLUSTERS, INNOVATION AND ENTREPRENEURSHIP – ISBN 978-92-64-04442-5 – © OECD 2009

institutions in the region that feeds graduates into the local talent pool. The majority of local high-tech firms require university educated employees, and in many cases, most of the staff has at least a B.Sc., many have M.Sc.'s, and a large number of firms have several staff members with a Ph.D., many in software engineering. While Waterloo is cited most often as the primary source of new hires, especially out of the software engineering programme, Wilfred Laurier University is regularly mentioned as a source for junior marketing and management people. Many firms, in both ICT manufacturing and software, have a mix of university-educated engineers and college-educated technicians, and report that they actively recruit from the local community college, Conestoga College, for their technical staff.

The influence of the local post-secondary institutions on the supply of highly educated and skilled workers in the labour force is identified consistently as a critical factor that drives the growth of the Waterloo cluster. Central to the education and training role that the University plays in the local economy is the co-operative educational programme that dates from the origins of the University. The University has the largest co-operative education programme in the world, with over 11 000 students (60 per cent of the student body) and 3 000 employers, 281 of them local, involved in the programme each year. Co-op programme offerings are extensive and are available in all faculties and departments, and over 100 different programmes. Many of the larger Waterloo firms, as well as global ones, have deep and enduring links with the co-op programme. At Sybase, an enterprise software company that spun-off from the original WATCOM Corporation, with over 250 employees in its Waterloo campus alone, 15 per cent of its current employees is Waterloo co-op students, and more than half of their Waterloo staff is former co-op students. Sybase also actively supports co-op activities at the high school level, and employees speak at local high schools, colleges, and universities about co-op education.

Three key benefits of the co-op programme were reported. First and foremost, it acts as a steady source of new hires, because firms know that the students have work experience, and they get the opportunity to evaluate them in the workplace before hiring them. Second, co-op students act as an important source of knowledge transfer; because they are exposed to new ideas in their courses and bring these ideas to their placements, "a lot of the students are on the cutting edge of the products that we're working on, so we definitely get the benefit from that". Finally, Waterloo co-op students have an international reputation for being of high quality, and as a result, local firms have to compete with global ones to attract the best students, though they retain the benefit of location. The growing international reputation of the University is reflected in the recently staged corporate recruiting drives by both Microsoft and Google, part of a broader upsurge in hiring within the region. A recent survey of 173 CEOs completed by Communitech reported that 80 per cent of the local firms are currently hiring, with more than half expecting to add 20 per cent to their workforces in the next year.

The multiple roles of local universities and colleges as R&D and tech transfer facilities, as well as suppliers of highly skilled talent, underscores the idea of the "embeddedness" of the local educational institutions in the Waterloo cluster. Regardless of whether firms had formal or informal links to the University, most of them cited the existence of local universities and colleges as a critical element of the cluster's dynamics. Some firms that are heavily networked with a local university describe a deeply synergistic relationship that has emerged and endures as a result of the university being located in Waterloo. Even firms with tangential or no ties to the university – for example, those who only hire co-op students or who comment simply on the international cachet of the University of Waterloo – cite the presence of the university as a critical factor.

Most of the firms in the Waterloo high-tech cluster are engaged in R&D. The Waterloo Tech Industry Profile reported that 76 per cent of the firms have R&D staff located in the region totalling 2 300 people. Large firms accounted for 51 per cent of the total number of R&D staff. In addition to in-house staff, 22 per cent of the firms reported that they also use external sources for R&D (Communitech, 2005, 11). However, there is a wide disparity in the R&D capabilities of large and small firms. Large firms typically have robust in-house R&D units, although even the small ones typically have some type of in-house development group that either focuses on a core idea to get it market ready, or engages in small, limited one-off collaborative projects, or "skunk works".

The firms in the region indicate that they are typically more focused on product development than exploratory research. The emphasis is predominantly on solutions-focused, incremental innovations rather than research-intensive, first generation innovations. Product and process improvements are intended to make the product "faster, smaller, cheaper" and often involve development activities such as the modification of existing software platforms, product updates and new releases, applying the core technology to different applications within the same factory, or making software web accessible. This emphasis on performance improvement and fine-tuning reflects the trend toward what one observer labelled "little R, big D". However, there was also evidence of robust R&D capacity reflected in the strategic decisions of large multinational firms, which often choose to augment their R&D or other technological capacity through the acquisition of local firms. Of particular interest is the fact that, while several large local firms have acquired foreign (primarily European) firms, several large foreign multinationals have acquired indigenous Waterloo firms – Google being among the most recent – to augment their R&D capabilities.

 CLUSTERS, INNOVATION AND ENTREPRENEURSHIP – ISBN 978-92-64-04442-5 – © OECD 2009

Large, global firms that collaborate with the University on long-term, core research projects, report that the primary benefit of their collaboration is "getting the first look" at research results. They want to keep abreast of what is happening at the research level, even though they know they will not have any proprietary access to the intellectual property (IP). Long-term research is by nature exploratory and speculative, and if they see it being directly relevant to the firm's business strategy, they prefer to keep the project within the company to avoid a potential conflict over ownership of IP. Ongoing involvement in university-based research also gives them an inside eye on developing university graduates who they may want to hire. Firms also report the benefit of research collaboration with the University as increasing their global reach and perspective by "magnifying your insight into the global marketplace", because research professors are usually part of global networks of expertise in their particular research areas. The majority of firms, both large and small, that report R&D linkages with a local university indicate that it is primarily for short-term research, usually of a couple months' duration, on a "project by project basis as needed" and that the primary benefit of collaboration is the ability to do problem-focused research and small co-development projects that allow them access to university expertise and lab facilities.

In contrast to the line of theorising about clusters that emphasises the importance of dense networking relationships among local firms, it is readily apparent from talking to software and ICT manufacturing firms in the Waterloo cluster that the amount of inter-firm collaboration in the form of key customer or supplier relationships is relatively low. The focus of most economic activity – key customers, sources of supply, competitors, and important strategic partnerships – for the vast majority of firms occurs at the continental and/or global level. While larger firms tend to be more focused on the global level, smaller firms, regardless of whether their key customers are currently in Canada, also have a growing global reach, or continental or global aspirations.

Many of the firms in the cluster describe co-location with customers, suppliers, or strategic partners as either unimportant or irrelevant. Of the firms for which proximity to customers is important, only very few have key local customers with whom they are in regular contact, and many firms treat local and non-local customers much the same, communicating primarily by phone or e-mail regardless of proximity. One firm commented that the distinction between local and global is very artificial. Most local firms have an explicitly global focus because for many, even their largest customer contributes only a small percentage of total revenue. This means that they have to "compete locally on a global basis", and find that they have to establish some type of local customer interface capabilities to serve their global customers. Customer relations – both marketing and support – occur at a virtual level. With "the Internet as the great equaliser", customers from all over the world can visit company websites to extract required downloads and access "24/7 customer support". Firms can choose to visit customers on site to deal with crises or complex issues, so proximity to customers is not a huge factor. Supplier relationships evince similar patterns as the vast majority of firms indicated that co-location with suppliers was not particularly important. Manufacturing firms tend to buy ready-made components, primarily from the US. While some firms do have local suppliers, they were not typically for key components, and when it was for a key component, the reason they sourced supplies locally was "because the type of technology they provide is more critical for us".

Role of SMEs

While there are some extremely large players in the area – Research in Motion (RIM), COM DEV, Open Text, AGFA, MKS, and Descartes Systems being the best known examples – most of the high-tech firms fall into the micro and small-sized enterprise category. Almost 70 per cent of high-tech firms in the Waterloo region employ between one and nine individuals, 20 per cent have 10-49 employees, and around 6 per cent fall into the 50-199 employee range. Only 3.6 per cent of the firms in the region employ over 200 people. Even more striking is the size distribution of establishments in both the ICT clusters – with 71 per cent of the establishments ranked as micro and small in manufacturing and 76 per cent ranked as micro in services. Unlike other concentrations of high-tech activity in Canada, the economy of the Waterloo region is not dominated by one particular sector, such as telecommunications or Internet-based firms. This diversity has enabled the regional economy to weather economic shocks – such as the post-2000 dot-com meltdown – that devastated employment in other leading ICT clusters across the country.

While firms may have some local partners who integrate their technology (or whose technology they integrate) into their product, key strategic partnerships, especially for larger firms, tend to occur overwhelmingly at the non-local level, and most often in the US and Europe. Key partnerships are often with a key customer or a key supplier because "your clients typically become your partners – your best sources of innovative ideas". Local partnerships, both formal and informal, tend to be rather weak and take the form of short, project-oriented collaborations, often on a contracting-out basis, with key local customers. These linkages tend to be "more relationships than formal alliances". From an analytical perspective, firm conceptions of what constitute a strategic partnership are inconsistent and the distinction between key suppliers or customers and strategic partners with whom they share common ownership or IP is not always clear, indicating that these

 CLUSTERS, INNOVATION AND ENTREPRENEURSHIP – ISBN 978-92-64-04442-5 – © OECD 2009

relationships are perhaps more fluid than is frequently understood. Local partnerships tend to occur primarily between smaller firms. Thus, the primary role of the SMEs in the cluster does not include the classical role of forming part of the supply base or architecture for the larger more globally oriented firms. The SMEs in the Waterloo cluster are typically outward looking themselves and although there is some evidence of sub-clustering at the local level, the relationships among SMEs are typically "soft" and informal.

It is rare, but there is some evidence of robust partnering activity between large globally oriented local firms and smaller ones. The relationships are both formal and informal, focused on taking up useful solutions that are generated by small local software firms, and are often initiated through common customers. One firm identified this as a key strength of the local cluster: "one of the values of being located here is that you have a lot of entrepreneurial small companies who are coming up with interesting solutions that we can attach to our offerings to round them out and bring them to market."

Impact of the cluster on entrepreneurship and employment

Since 1976, the number of high-technology enterprises in the area has grown to a critical mass, starting with a flurry of new firm formation that included a combination of spin-offs from the university, in-migration of firms from outside the region and independent start-ups. Of these sources, university spin-offs have had the greatest impact on the local economy. The University of Waterloo is among the best performing universities in Canada in terms of the number of spin-off companies it has produced. Since 1973, the University of Waterloo has spun off 59 individual high-technology firms, 28 per cent of the total number of high-tech firms born in the cluster (Xu, 2003). Some of the most notable spin-offs include Waterloo Maple Inc (1988), Open Text (1989), Virtek Vision Corp. (1986), Dalsa (1980) and Northern Digital Inc (1981). The University of Waterloo's Technology Transfer and Licensing Office identified 106 spin-off companies employing over 2000 people by the mid-1990s.

Using a somewhat different definition that included the transfer of intellectual resources, the PriceWaterhouseCoopers' study of regional economic benefits identified over 250 spin-off companies from the university (2001). Independent start-ups and second and third generation spin-offs also contributed greatly to the high-tech growth in this period. The recent Waterloo Region Tech Industry Profile reported that 52 per cent of executive respondents were alumni of local colleges or universities, with the majority coming from the University of Waterloo. In addition, 70 per cent were the founders of their own companies, with 31 per cent of these being serial entrepreneurs who had created and sold a number of companies

(Communitech, 2005, p. 13). Much of the university's commercialisation and spin-off success is attributed to its IP policy, which allows ownership of IP to rest with the creator, thus encouraging the individual (faculty or student) to commercialise the idea.

The Waterloo ICT cluster is thus distinguished from some of the leading ICT clusters in North America by the relative predominance of its small and medium-sized enterprises. However, the growth of the cluster can be attributed to the rapid expansion of some of its leading firms, particularly Research in Motion, as well as the slow, but steady increase in the number of smaller new firms. One minor, but significant issue for the cluster is balancing the needs and demands of the leading firms, for office and manufacturing space as well as personnel, while ensuring the adequacy of supply for the smaller firms that have less market power.

The promotion of entrepreneurship among the SMEs has not historically been a central focus of policy at either the national or provincial level. However, a number of more clearly local factors have contributed to enhancing the entrepreneurial skills and capacity of local SME owners and managers. The region is home to a relatively dense network of local business and industry associations that have demonstrated the ability to work together to promote the interests of the local cluster and provide a strong degree of mutual support to each other. The regional culture in the region is characterised by a robust "entrepreneurial spirit" supported by a small and transparent business community and well-developed business associations, as well as a vibrant social network and sense of community. This type of associational activity is evident in the growth of regional associations in the Waterloo high-tech community focused on facilitating the region's economic competitiveness and sustainability. Canada's Technology Triangle (CTT), the Communitech Technology Association, the local Accelerate Network (now part of Communitech) and the Waterloo Region Prosperity Council all play important roles in supporting regional economic development. Communitech, formed in the late 1990s to lobby the government in the interests of high-technology business, has been an important addition to the institutional infrastructure of the cluster in the Waterloo region. It was created as an initiative of a group of high-tech entrepreneurs with the specific purpose of establishing cutting edge infrastructure to support regional high-tech prosperity, expansion and global competitiveness. An often-cited benefit of Communitech membership is access to a pool of shared experiences and support through seminars, Peer2Peer sessions, networking events, and conferences. More recently, CTT, Communitech, the Greater Kitchener-Waterloo Chamber of Commerce and the Cambridge Chamber of Commerce have come together as the Prosperity Council of Waterloo region to collectively create an environment that supports opportunities for prosperity in Waterloo region. Together they represent more than 3 000 businesses in Waterloo region. Prosperity, for the purposes of the Council, involves initiatives and policies that support wealth creation, supporting the objectives of enhancing the standard of living and overall quality of life in the region.

The University of Waterloo is also active in its support of entrepreneurial education and activities. The mandate of the recently established Centre for Business, Entrepreneurship and Technology (CBET) is to co-ordinate, develop, and support the several strands of UW's entrepreneurship activities, all of which are intended to facilitate the development of UW as an "Entrepreneurial University". More specifically, CBET is intended to research issues such as "how an entrepreneurial culture is created within a university, how faculty members commercialise their technology, issues of the relationship between academic researchers and the business community and issues relating to the impediments of facilitating a transfer of technology between those two communities". In terms of educational programmes, it has recently launched the Master of Business, Entrepreneurship and Technology (MBET), which attracts potential entrepreneurs from around the world, and teaches business skills critical to identifying, exploiting, and establishing new commercial opportunities, with an emphasis on innovative technologies. Undergraduate students can also participate in the Enterprise Co-op programme where they commercialise a business venture of their own rather than work for an existing firm.* Innovate Inc. is a department within the university that provides resources and counselling to faculty and student entrepreneurs, and aims to facilitate the commercialisation of knowledge created within the institution. Finally, the Institute for Innovation Research, affiliated with the Faculty of Engineering, is dedicated to the generation and dissemination of applied interdisciplinary research that advances understanding of entrepreneurship in technology-based enterprises, and to promoting entrepreneurship within universities.

Barriers to cluster development

Although the cluster in particular, and the Waterloo region more generally, are currently viewed as doing quite well, informed observers acknowledge that the region faces a number of critical challenges. One of the key challenges identified by a number of insiders is the fact that many of the leading firms today were founded in the late 1970s or 1980s, but that the pace of new firm formation has fallen off significantly in the past decade and a half.

* According to a university official, in the Enterprise Co-op Program, "a small number of students are encouraged to start their own companies during co-op work terms" and "we take about 10% a year of those people who think they've got it and we give them a very rough screening process where we explain to them that this will be the toughest co-op term that they ever have had. We give them a small amount of funding, somewhere between CAD 6 000 and CAD 8 000, and then we mentor them."

This has raised concerns within the cluster about its internal ability to continue to grow and expand. Closely related to this is the prominent role played by several leading firms in the cluster, such as RIM or Open Text that specialise in particular niche markets. While these firms are currently enjoying great success – RIM with the recent launch of the Blackberry Pearl device and Open Text with the acquisition of one of its leading competitors, Toronto-based Hummingbird, – they are vulnerable to sudden shifts in market demand or the emergence of new, unanticipated competitors. To date, this has not occurred and the future of the cluster looks secure, but it is a potential source of concern.

Another challenge that reflects the opposite side of the coin is coping with the cluster's current level of success. In a recent interview with the local high-tech association, the major issue that was identified was a looming shortage of software engineers due to the high projected level of hiring by cluster-based firms. The recent establishment of a major research facility by Google within the Waterloo region and their announced plans for future hiring, the continuing expansion of RIM within the Waterloo region and the fact that Microsoft views the Waterloo region, and the University of Waterloo in particular, as a major source of recruiting for its Redmond, Washington operations are placing considerable pressure on the ability of both the local post-secondary institutions to meet the demand for highly qualified personnel. The ability of the cluster to satisfy this increasing demand for personnel will be critical for its continued success; as noted above, the strength and "thickness" of the local labour market has been a key driver of its competitive success to date.

Somewhat less threatening, but still of concern, are a number of infrastructural issues that hover in the background of the cluster. While Waterloo is a major urban centre in southern Ontario, it does not enjoy direct transportation connections, especially by air, on it is own. It is located less than an hour's drive west of the Toronto (Pearson) International Airport, upon which it relies for international flight connections. However, the highway corridor connecting the Greater Toronto Area with the Waterloo region (Highway 401) is also the major transportation route for Ontario's substantial trade, especially in automobiles and auto parts, with the US Midwest. Traffic congestion in this corridor is becoming an increasing problem for the entire southern Ontario economy and represents a growing issue for firms based in the Waterloo region. This is a major challenge faced by the region in general, and the ICT cluster in particular, arising from the substantial urban sprawl occurring in the larger Greater Toronto Area and the spillover effect it has on the Waterloo region.

A final issue concerns the lack of political integration among the various municipalities that comprise the Waterloo region – particularly, Cambridge, Kitchener, Waterloo and Guelph. A province-wide process of amalgamating local municipalities in the late 1990s encountered major opposition in this

 CLUSTERS, INNOVATION AND ENTREPRENEURSHIP – ISBN 978-92-64-04442-5 – © OECD 2009

region and the proposed amalgamation was not imposed on the area. In addition to the four local municipal governments, there exist a number of county administrations in the rural area surrounding the municipalities, as well as a regional level of government superimposed on the municipal ones. The result is a certain fragmentation of administrative responsibility for issues such as region-wide transportation planning or local economic development and business attraction. It also results in a certain amount of duplication of government offices across the many separate jurisdictions in the region. Many of the business leaders in the cluster recognise this political fragmentation as a drawback for the region, but believe that the continuing strength and vitality of the local economy has generated sufficient positive feeling in the region to encourage the local governments to co-operate and work effectively together on common interests.

Role of policy

The case of the Waterloo ICT cluster demonstrates the long-term impact of expanding research infrastructure on the development of local clusters. However, it is important to recognise that, to a large extent, the cluster emerged as an indirect, and partly unintended, consequence of policies that were directed towards meeting other goals and objectives. The cluster concept has only become of interest to policy makers at all levels of government since the early 1990s. As the preceding narrative makes clear, the roots of the cluster lie more than forty years in the past, long before the current period of interest in this phenomenon as a policy instrument.

The case study also raises another issue that confuses many analyses of the origins of clusters – namely, the respective role of different scales of political jurisdiction in the genesis of clusters. While clusters are primarily seen as key features of local and regional economies, and most of the literature and case studies highlight the contribution made by local factors and industrial dynamics, the presence of the senior levels of government lurks in the background. A number of studies highlight the relationship between the cluster concept and others used to analyse the innovative capacity of regional and national economies, principally the innovation systems approach. The concept of "nested scales" describes the interacting set of effects that different levels of government exert on firms operating in a specific geographic location. From this perspective, clusters are seen as nested within, and impacted by, other spatial scales of governance, including regional and national innovation systems, each of which adds an important dimension, or layer, to the economic structures and government policies that impact firms within the cluster.

Various elements of each of these spatial scales may have significant effects on the innovation process and competitive dynamics within the cluster. For instance, the national innovation system may play a preponderant role in establishing the broad framework for research and innovation policies, in providing a national system of research organisations, in establishing the rules of corporate governance that influence firm behaviour, in setting the rules of operation for the financial system that determine the availability of different sources of financing and time horizons for new and established firms, and finally, for setting the broad framework for the industrial relations, employment and training systems that influence job paths, inter-firm mobility and skill levels for the labour force. Levels of regional specialisation as encompassed in the concept of regional innovation systems play an important role in affecting cluster performance through the provision of the regional/state/provincial research infrastructure, specialised training systems, the broad education system, policies for physical infrastructure and the investment attraction function. At the local level, high levels of civic associationalism, particularly the business-higher education link, can exert a strong influence over cluster development. The local level also plays an important role in the provision of infrastructure, such as roads and communication links, as well as in the governance of the primary and secondary education system, which are significant factors for cluster-based firms in the attraction and retention of managerial talent.

In the case of the Waterloo cluster, the policies that exerted the greatest impact on the development of the cluster have been federal and provincial policies in support of post-secondary education more generally, and research funding, in particular. In Canada the provinces exercise primary responsibility for funding post-secondary education, but there have been broad cost-sharing agreements in place with the federal government since the 1960s. The precise mechanisms used to provide for the cost sharing blend elements of both revenue and expenditure policy instruments. In addition, the federal government has primary responsibility for funding and supporting post-secondary research in Canada through the three federal granting councils, but since the 1980s, this funding support has been augmented and reinforced by a growing number of provincial programmes. In the late 1990s the federal government created the Canada Foundation for Innovation (CFI) to fund the updating of research infrastructure at post-secondary institutions across the country, as well as the Canada Research Chairs (CRC) programme to create 2000 federally funded research chairs at universities (Wolfe, 2005). The increased focus of the province on providing direct support for research was underlined with the creation of a distinct Ministry of Research and Innovation in 2005, with the Premier of the Province as its Minister.

In addition, both the federal and provincial governments offer a wide array of other policies of benefit to individual firms in the cluster. The federal Scientific Research and Expenditure Development (SR&ED) Tax Credit, combined with provincial R&D incentives, offers one of the most generous tax

 CLUSTERS, INNOVATION AND ENTREPRENEURSHIP – ISBN 978-92-64-04442-5 – © OECD 2009

jurisdictions in North America for the conduct of research. Other federal programmes, such as Technology Partnerships Canada (recently suspended by the new federal government) have provided direct subsidies to firms conducting innovative research. One of the most widely accessed federal programmes is the technology adoption support provided through the Industrial Research Assistance Program (IRAP), administered by the National Research Council. The local IRAP representatives, or Industrial Technology Advisors (ITAs), in Waterloo work closely with the regional high-technology association, with their offices co-located with Communitech and the Accelerator. The provincial government, both prior to and continuing with the establishment of the Ministry of Research and Innovation has dramatically expanded its support for research funding over the course of the past decade – partly to provide matching funding for some of the federal programmes. More recently it introduced a number of key programmes and initiatives designed to accelerate technology commercialisation across the province with a focus around the new MaRS commercialisation centre located in the heart of the university and research hospital district of downtown Toronto (Wolfe, 2006).

Policy adaptations over time

As the preceding discussion of the evolutionary path of the Waterloo ICT cluster suggests, the origins of the cluster owe much more to the broad framework of federal and provincial policies supporting research and education than it does to the direct effect of policies explicitly designed to stimulate cluster development. None of these federal or provincial programmes were specifically designated as cluster initiatives or targeted at cluster promotion. The positive contribution they make to clustering at the local and regional level is a welcome, but indirect, and sometimes even unintended, consequence of their explicit programme goals or objectives (Wolfe and Gertler, 2006).

However, as the cluster has grown and developed, a number of recent initiatives have been directed at providing increased support for the firms in the cluster, as well as accelerating the pace of technology commercialisation and new firm spin-offs from the local universities. Both the federal and provincial governments have adopted limited cluster policy initiatives over the course of the past decade. Canada's National Research Council (NRC) has pursued an explicit strategy of developing clusters around several of its research institutes since the late 1990s. The strategy involves the deliberate effort to transfer technology out of its newly established research institutes and promote the growth of a cluster of related firms in the regional economy around the institute (OECD, 2007). The strategy has been applied to fourteen NRC institutes across the country, but there is no NRC institute in Waterloo and therefore the cluster development strategy has not been implemented in this region. In its Innovation Strategy, released in 2002, the federal government expanded on this approach in announcing the objective of developing ten internationally competitive clusters across the country, but this strategy was never fully implemented and the objective has not been endorsed by the current federal government.

A more recent policy initiative launched by the Ontario government, the Biotechnology Clusters Innovation Program (BCIP) was directed towards the explicit goal of cluster promotion. The provincial Minister of Innovation launched Ontario's Biotechnology Strategy in June 2002. As part of that strategy, the government announced a new programme initiative: the Biotechnology Cluster Innovation Program (BCIP) with the goal of accelerating the development of Ontario's biotechnology clusters by supporting the commercialisation of infrastructure projects and the diffusion of biotechnology-related innovations into knowledge-based or traditional industry sectors. The programme consisted of two distinct phases. In the first phase, the government supported the development of plans that address the innovation capacity of Ontario's regional biotechnology clusters. The programme provided funding up to a maximum of CAD 200 000 on a matching basis, to regional consortia for the development of a Biotechnology Cluster Innovation Plan, including one in the Waterloo region. The second phase of the programme was designed to support the development of infrastructure such as commercialisation centres, research parks and other regional initiatives that promote entrepreneurship and innovation. Eleven regional consortia developed regional innovation profiles and corresponding regional cluster strategies in the first phase of the programme. Between late 2003 and early 2005, provincial officials held a series of seminars with representatives of the eleven consortia, as well as separate meetings with the individual groups. The original BCIP strategy developed for the Waterloo Region had little direct involvement from the ICT cluster or significance for it, given that the overwhelming focus of the initiative was on biotechnology. However, this changed in 2005 with the refocusing of the programme on a broader cross-section of industrial sectors and clusters.

In the provincial budget of May, 2005, the government launched the follow on phase of the programme in the form of a series of "regional innovation networks" (RIN). These are described as "multi-stake holder, regional development organisations established with provincial funding that support partnerships among business, institutions and local governments to promote innovation". The regional innovation networks are mandated to expand beyond their original focus on the life sciences to include other areas of innovation excellence, such as information technology, energy conservation and advanced materials, depending on their local strengths and opportunities. The networks are also described as constituting part of a multilayer commercialisation

 CLUSTERS, INNOVATION AND ENTREPRENEURSHIP – ISBN 978-92-64-04442-5 – © OECD 2009

network that includes the province, multiregional groups focused on key technology areas or industrial sectors and the original regional consortia, described above. The constituent parts of the network are to support two complementary sets of activities – those that build on and connect the components of the network and those that contribute to a more effective alignment of existing federal, provincial and local research infrastructure and related innovation assets. A key function of the networks is to increase the knowledge flow and build linkages between existing research institutions and firms in order to build industrial capacity for the uptake and adoption of new research and technology. The overriding goal of the RIN's is to increase regional innovation capacity by addressing commercialisation gaps in the existing level of support for small and medium-sized enterprises in regional networks across the province in innovation-intensive sectors and clusters, including Waterloo. The programme also aims to develop strong networks that can increase the accessibility of the public research infrastructure and resources for firms (Wolfe, 2006). While the transition from the earlier BCIP programme to the RIN's is still in its early stages, overall, the programme displays many of the positive features of bottom-up strategic planning. In the Waterloo context, the local RIN is building upon the cluster's past success, the existing strength of the post-secondary research infrastructure and recent new initiatives, such as the Waterloo Research and Technology Park (see below), to accelerate the commercialisation of new research results and sustain the process of new firm formation and growth within the cluster.

A related initiative designed specifically for the Waterloo region has been the establishment and marketing of the new Research and Technology Park. Support for the Research Park has taken the form of a partnership between the University of Waterloo, the Government of Canada, Province of Ontario, the Region of Waterloo, the City of Waterloo, and Communitech. The new Research Park is located on the northern boundary of the University of Waterloo and is already the site of several new buildings, one of which houses Communitech, the Accelerator Centre, the local IRAP offices, legal offices and other support functions directed towards the promotion of local firms in the cluster. Other buildings, both completed and currently under construction, will house some of the leading firms in the cluster or provide space for growing firms to expand into, as well as a number of amenities deemed to be attractive to employees in high-tech firms. The co-operative role played by all major levels of government plus the key cluster actors in designing, financing and developing the new Research and Technology Park is strong testament to the importance they all attach to the further growth of the cluster. A number of strategic planning exercises launched through the Prosperity Council and involving the key industry associations in the region are also directed towards supporting the growth of the cluster.

Future policy challenges

The crucial implications to be drawn from the preceding analysis are the necessity of developing a rich and thick labour market in the skills required to build the industries and technologies in the cluster, then to support the interaction within the cluster through the nurturing of strong social networks among government partners and industrial leaders, thus building a sense of civic engagement. This has worked well to date largely through the unplanned but highly effective interaction between the federal and provincial governments, strong support for building the local post-secondary research infrastructure in the region and providing increased research funding over the past ten years to those institutions.

The cluster as it is currently constituted faces a number of critical policy challenges. As noted above, the rapid pace of new firm formation in the region was stronger in the 1980s and early 1990s than it has been over the past decade and a half. Despite the increase in the number of experienced entrepreneurs in the region, some of whom devote considerable time to acting as angel investors and mentors to new firms, many key civic and business leaders in the region express concern over this recent decline in the pace of new firm formation. Given that the preceding pace of firm formation was largely unplanned, it is not clear that this deficiency can be remedied purely by policy measures. The recent expansion of the University of Waterloo's efforts in promoting entrepreneurship teaching and research, and the activities of initiatives such as the Enterprise Co-op programme and Innovate, Inc. within the university may help remedy this deficiency but it is still too early to tell. In addition, as noted above, one of the most effective mechanisms for providing these kinds of supports are through the peer to peer mentoring networks established and maintained by the local high-technology industry association, Communitech. Government policy could be designed to provide more direct and explicit support for these initiatives than it has in the past. There is also considerable scope for more effectively integrating the activities of the various local industry associations, discussed above, into the activities of the Waterloo Region RIN.

Another current policy challenge facing the cluster is the recent shortage of a sufficient number of graduates to meet the perceived demand for future hiring, both as a consequence of the internal expansion of leading firms, such as RIM, as well as the inward location of global firms, such as Google. Expansion of university programmes in the areas of high demand would appear to be the most obvious solution. This was done in the late 1990s through the provincial Access to Opportunities Program that aimed to create 51 000 new graduates in computer-related programmes, but there is some concern that recruitment of sufficient applicants into such programmes might prove more challenging than it was a decade ago.

Lessons for other clusters

The factors which contribute to the emergence of clusters in the first place and sustain the dynamism and growth of cluster-based firms remain of critical interest. The case of the Waterloo ICT cluster sheds light on several key issues related to the core question of location: if co-location with other firms is not a main driver of local economic growth and innovation, what accounts for the resilience or "stickiness" of the cluster? Local economic growth in the Waterloo cluster is a result of the interaction between location, institutions and a regional entrepreneurial culture. Whereas purely locational factors, based on demanding local customers, suppliers and competitors, do not fully account for the emergence or growth of the Waterloo cluster, the presence of key local institutions – primarily the University of Waterloo and a dense network of local civic associations – provide the glue that retains and sustains innovative high-tech firms.

The case of Waterloo Region's ICT cluster also provides support for the argument about the long-term impact of expanding research infrastructure on the development of local clusters. In the Waterloo case, the mobilisation by local business leaders to secure a charter for a new university, financed with federal and provincial funding, and their foresightedness in structuring a curriculum around math, sciences and engineering and creating a pioneering programme of co-operative education, all laid the groundwork for the future emergence of a dynamic and growing information technology cluster. However, one should not overlook the supportive role played by the senior levels of government whose expansion of support for the post-secondary education system in the 1960s made possible the establishment of a new university. It was the specific pattern of interaction between dynamic, visionary leaders at the community level with the increase in federal and provincial funding that created the local antecedents essential for the emergence of the information technology cluster (Wolfe and Gertler, 2006).

Whether intentional or inadvertent, one of the most effective public policies for seeding cluster development is a sound investment in building the research and skilled labour base in a region. The establishment of a strong local talent pool of highly skilled and knowledgeable workers both feeds the growth of the local firms in the cluster as increasing returns begin to take hold, and attracts outside firms to locate in the cluster in order to gain access to the local knowledge-base and the skills embedded in the local labour market. However, the presence of a strong local research infrastructure and a "thick" local labour market may not be sufficient on their own to spur the formation of a local cluster. The Waterloo ICT cluster owes its current success to the effective intersection of a strong sense of civic engagement with the rich knowledge resources afforded by its strong research infrastructure and talented local labour market.

The findings from this case study in Waterloo echo the findings from cluster studies in other regions and economic sectors across Canada, and suggest a different conception of what factors really matter in cluster formation and sustenance than some of the dominant theories in the literature. Concerning the dynamic relationship between local firms in the cluster, the findings indicate that for many of the most successful clusters the most important end-user markets and knowledge networks are continental or global (Wolfe, Davis and Lucas, 2005). The most significant local factor is a "thick" labour market that provides a steady supply of the highly skilled personnel for firms to draw upon. This factor combined with the development of strong social networks and a strong sense of civic engagement is crucial.

References

Canada's Technology Triangle (2004), *Community and Statistical Profile,* Waterloo.

Communitech Technology Association (2005), *Waterloo Region and Guelph Tech Industry Profile: 2005 Survey Results,* Waterloo, Ontario.

Gertler, Meric S. and David A. Wolfe (2004), "Local Social Knowledge Management: Community Actors, Institutions and Multilevel Governance in Regional Foresight Exercises", *Futures,* Vol. 36, No. 1 (February), pp. 45-65.

OECD (2007), *Competitive Regional Clusters: National Policy Approaches,* OECD reviews of Regional Innovation, OECD, Paris.

PriceWaterhouseCoopers (2001), *University of Waterloo: Regional Economic Benefits Study,* Waterloo.

Wolfe, David A. (2005), "Innovation and Research Funding: The Role of Government Support", in F. Iacobucci and C. Tuohy (eds.), *Taking Public Universities Seriously,* University of Toronto Press, Toronto.

Wolfe, David A. (2006), "The Role of Higher Education and New Forms of Governance in Economic Development: The Ontario Case," in *How Universities Can Promote Growth,* Shahid Yusuf (ed.), The World Bank, Washington, D.C.

Wolfe, David A., Charles Davis and Matthew Lucas (2005), "Global Networks and Local Linkages: an Introduction", in David A. Wolfe and Matthew Lucas (eds.), *Global Networks and Local Linkages: The Paradox of Cluster Development in an Open Economy,* McGill-Queen's University Press, Montreal and Kingston.

Wolfe, David A. and Meric S. Gertler (2006), "Local Antecedents and Trigger Events: Policy Implications of Path Dependence for Cluster Formation", in P. Braunerheim and M. Feldman (eds.), *Cluster Genesis: The Emergence of Technology Clusters,* University Press, Oxford.

Xu, Stephen Xingang (2003), "Knowledge Transfer, Interfirm Networking and Collective Learning in High Technology Cluster Evolution: A Network Analysis of Canada's Technology Triangle", Master of Applied Science (Management Science) dissertation, University of Waterloo.

CLUSTERS, INNOVATION AND ENTREPRENEURSHIP – ISBN 978-92-64-04442-5 – © OECD 2009

ISBN 978-92-64-04442-5
Clusters, Innovation and Entrepreneurship

Chapter 9

Conclusions and Recommendations

by

Gabriela Miranda and Jonathan Potter
Centre for Entrepreneurship, SMEs and Local Development, OECD

This chapter aims to identify key messages from the book for the definition of economic development and job creation policies suited to the context of geographical concentrations of related industries or "clusters", drawing on the experience of already validated strategies and practical models.

The chapter outlines the factors which explain the development of successful clusters and the challenges to be faced in their development. In the second part, we present recommendations to support and strengthen the economic impact of clusters. These recommendations apply to all the players who have an influence on the cluster at national, regional and local levels, both in the private and public sector. In the last part, we show how these recommendations have been implemented by the various players in the clusters studied. We are talking here of "good practice" that can be adopted by other clusters elsewhere. For ease of reading and understanding, the recommendations and good practices are reproduced in the form of tables at the end of the chapter.

Clusters in the global knowledge economy

To boost economic development and respond to ever keener international competition, the governments of OECD member countries need to pursue entrepreneurship and innovation policies that reflect not just their own national institutional and economic environments but also the distinctive needs of local entreprise, research and training "clusters". These clusters create an environment conducive to productivity gains, which are a key factor of growth, and so form structures that help meet the challenges of international competition. Nevertheless, public policy can often play an important role in strengthening clusters.

The case studies analysed showed the different approaches and initiatives adopted both nationally and at regional or local level. Even if the development of clusters and their local conditions (geographical area, society, environment, etc.) cannot be generalised, we present here a list of conditions which seem more generally to foster or hinder the development of clusters and the scope of their economic impact.

Key factors in cluster success

Networks, innovation and the supply chain

Clusters are often associated with a strong fabric of small and medium-sized enterprises (SMEs), which often draw on collaborations with a few key large enterprises and/or universities and research organisations for their innovation and competitiveness. These collaborations range from simple participation in supply chains to intensive co-operation on innovation projects. Such networking provides the opportunities for identifying customers and sharing information, thus helping build supply chains. Although these are private transactions and strong collaborative environments often arise without public intervention, the public sector has also played a role in creating the conditions for collaboration in certain cases, particularly through financing of collaborative research and development initiatives. However, governments could do more to stimulate bottom-up networks. In particular, public, private or part public part private cluster management structures can play an important role in promoting collaboration and the creation of formal networks both within and external to the cluster.

Leading enterprises

The data presented for several of the case study clusters indicate a considerable growth in the number of new and small firms and employment. The effect partly involves the emergence and growth of high-tech and other leading enterprises in the clusters but it is also partly indirect, through linkage and multiplier effects resulting from the purchases of firms and employees in the core enterprises of the clusters, stimulating growth in a range of suppliers to producers and consumers. Often, a few key enterprises account for the bulk of the expansion of high-tech enterprises in the cluster.

Strong universities and research centres

Another element essential for the emergence and growth of clusters is the presence of universities and research centres of international repute. Leading universities in their specialist fields and top research centres create knowledge that can be transferred to other actors in the local economy through spin-out companies, research collaborations, consultancy and informal contacts. They also create a critical mass of human capital, the cornerstone of the development of clusters.

Investment by the public sector

Public actors have often played a critical role in defining and implementing programmes or initiatives in support of successful clusters and this works best when there are strong mechanisms for partnerships between the State, regional authorities, business support agencies and local firms and research organisations. National, and sometimes regional and local, actors have played a significant role in the development of successful clusters through lending support to the financing of research infrastructure and research projects in particular as well as supporting universities and public research institutes. This support needs to be sustained over time and responsive to changes in markets and technologies. Many other initiatives and programmes developed by a range of business support organisations are also important, such as inward investment attraction, small firm support and training policies.

Quality of life

A strong critical mass of highly skilled and entrepreneurial labour is at the heart of the emergence and development of successful clusters. This in turn is related to the provision of a favourable environment such that the human capital stays in the area and grows. Social networks, people-oriented services, communications and infrastructure are just some of the elements which have an impact on the attraction and retention of the highly skilled population. A high quality of life is thus an essential factor in ensuring that the area is attractive and in encouraging the development of clusters.

Social capital

Finally, a factor which can be seen as fundamental in all the clusters analysed is the presence in the area of a strong social fabric. Social capital is recognised as a key factor in exchanges of ideas and innovation and, therefore, for clusters. Social networks and trust bring about the emergence and development of innovation by creating an environment open to exchange of information (formal or informal), capacities and knowledge which create value within the cluster and thus enhance its size and standing on the international scene.

Obstacles to the development of clusters and innovation

We have seen a brief overview of the conditions which seem to have a positive impact on the emergence and development of successful clusters. A strong case can be made that policies should seek to reinforce or replicate these factors in other clusters. However, in order to support clusters better and target that support effectively, a clear vision of the obstacles to be overcome is also needed. In this part, we present the obstacles to the further growth and development of the seven clusters studied. We concentrate on those which are faced by several of the clusters and that may affect cluster development more generally. This will allow us subsequently to put forward pertinent recommendations on how to overcome the obstacles which can be adopted by other clusters in other OECD countries and elsewhere.

Inadequate incentives to the commercialisation of public research

Much of the capacity for the continued successful evolution of established clusters lies in their capacity to create new enterprises based on new technologies, products and markets. Public research organisations are a major potential source of such start-ups. However, it has been found that spin-outs from universities are not always easy, even in established and successful clusters. There are often institutional barriers to researchers and academics to create enterprises. A weak culture of entrepreneurship is also often present in public research, which does not encourage those with ideas to leave and create their own enterprise. In addition, public support mechanisms for the expansion and creation of start-ups are often no stronger in the clusters than in the country as a whole, although it can be argued that stronger and more tailored start-up support should be available in these places. The problem of inadequate incentives for research commercialisation also concerns incentives for engaging in other types of commercialisation activities, including consultancy and collaborative research for industry. Although the issue appears to be primarily one of academic and research cultures that are unfriendly to entrepreneurial engagement, further problems lie in the low weight given to collaboration with industry in researcher performance assessments of researchers, unfavourable regulations governing the rights of researchers to use the intellectual property they help develop and difficulties for researchers to return to their jobs following periods spent in business. Commercialisation is also hindered by limited capacity of the public research sector to promote the products of innovation in the market. However, universities can play an important role in marketing if they have an active intellectual property policy.

Maintaining a critical mass of qualified labour

An important issue for several clusters concerns the training of qualified staff capable of anticipating and responding to changes in markets and technologies, covering not just the highly skilled such as researchers and professionals, but also skilled staff, for example laboratory assistants and operators of sophisticated machinery. To ensure the growth of clusters in the long-term, it is necessary to create mechanisms such that the human capital constantly adjusts to the new needs. This is particularly important in occupations and sectors with expanding demand. Several of the clusters face skills shortages in new activities. This requires both adequate training programmes, which may need to be co-ordinated or supported by the public sector and an infrastructure and quality of life to attract labour from other areas.

Lack of seed capital

Many of the chapters highlight the need for seed capital to support the development of innovative products and new and small enterprises. However, in the majority of the cases analysed, one of the chief problems is the lack of venture capital and other sources of private or mixed financing to meet the needs of advanced technology development, especially at the start-up phase and until production is under way.

Lack of a co-ordinated policy strategy for the cluster

A large number of public organisations typically have a role to play in various aspects of the development of the clusters studied. However, often their activities are not well co-ordinated leading to problems in developing a comprehensive and forward-looking strategy bringing to bear the resources of all the available actors in addressing current and future problems in cluster development. The need for improvements in policy co-ordination concern not just the range of actors involved in the local area but also the need for greater involvement of national governments in providing sufficient and tailored support for the evolution and growth of major clusters, which of course have economic benefits not just for the local area but also for the nation as a whole. As well as effective coordination of public policies based on a common strategy it is important to strengthen public-private partnerships and involve the private sector in initiatives to develop clusters, both in terms of their knowledge of cluster needs and in terms of the financial and other resources they can offer the development process.

Congestion

The rapid recent economic growth of the studied clusters has in some cases led to significant problems of congestion in their host agglomerations that may impede further growth, particularly where the cluster dominates its agglomeration and there is little capacity to expand the agglomeration because of planning or physical constraints. Strong growth in constrained agglomerations is associated with rising land and property prices, rising wage costs, increasing commuting times and transport problems that may reduce the attractiveness of the cluster to firms and labour and hence the ability of the cluster to generate further agglomeration benefits.

Risk of "social divisions"

The growth of clusters may also give rise to an increase in internal social inequalities within agglomerations between "high-tech" employees and other strata of the population. This can stem not only from disparities in incomes and levels of activity of the various groups in the labour market, but also from increases in the cost of living caused by the spending power of the new arrivals. "Social divisions" could undermine support for the cluster by local authorities.

Ethical objections

A further issue that has arisen in some clusters is that of ethical objections by members of the public to the nature of the research being undertaken – the ethics of the practices used and the applications the research results could be put to. Opposition has been encountered for example to research activities in the fields of nanotechnology and life sciences. It comes in the form of demonstrations, direct action and opposition by political parties. This may put into question public investment and policy support for cluster activities.

Recommendations

The above conclusions and analyses are an attempt to give a general overview of the dynamics of clusters and the challenges they face. Actors in the policy system that supports clusters have an important role to play, both in building strengths and addressing obstacles. All the players who support clusters, nationally and in regions, and in the public and private sectors, are concerned. To guide policy development, a series of recommendations on policies and initiatives is presented in Box 9.1. These recommendations are not exhaustive but they already offer a starting point for appropriate reforms. Each of the recommendations is discussed in more detail below. In addition, Table 9.1 gives examples of programmes undertaken in the case study clusters that illustrate how recommendations can be implemented in practice.

Encouraging entrepreneurship

Supporting spin-outs

Familiarising researchers with creation. Enterprise creation or merely bringing researchers closer to the world of industry can be a very positive step. Courses in "business plans" aimed specifically at researchers and similar programmes could be run in universities to encourage spin-outs, as is the case of Vienna.

Financing areas of incubation and spin-outs. It would be worthwhile to create areas of incubation with advice and assistance at regional level specifically to the academic world on ways of converting a good idea into a viable enterprise. Two examples are the initiatives taken at Medicon Valley (Bioincubator) and Madison (URP).

Box 9.1. **Summary of recommendations**

Encouraging entrepreneurship

- Supporting spin-outs.
- Leading a transition to the entrepreneurial university.
- Supporting the launch and growth of start-ups.

Stimulating innovation and collaboration

- Fostering SME-research collaboration.
- Encouraging collaboration: within the cluster and between clusters.
- Encouraging enterprise networks.
- Better marketing of products.

Coordinating public polices and regional initiatives

- Strengthening public-public and public-private partnerships.
- Encouraging evolution in cluster activities.

Ensuring quality human capital

- Updating education and training to meet the requirements of the cluster.
- Ensuring availability of talent locally.
- Ensuring the appeal of the area and a good quality of life.

Facilitating access to financing

- Encouraging private investment.
- Facilitating access to public funding.
- Creating forums to seek financing.

Reducing congestion and social divisions

- Addressing congestion and social inequalities resulting from the emergence of the cluster.
- Creating mechanisms to inform the community about the activities of the cluster.

Establishing formal collaboration projects. Clusters can encourage spin-outs and lead to considerable filtering down of technology to SMEs through collaboration between large enterprises, research institutions and SMEs, especially through relations with suppliers and subcontractors. Even though such technology transfers can occur spontaneously, in some cases, such as Vienna and Waterloo, they were encouraged by formal collaboration projects.

Leading a transition to the entrepreneurial university

Promoting entrepreneurial education. It is important to promote entrepreneurship in universities across a wide range of subjects, researchers and students. The entrepreneurial university must therefore be promoted, laying emphasis on issues relating to the capacity to create and market innovations, including teaching of key skills such as identifying, exploiting and creating new opportunities.

Supporting the launch and growth of start-ups

Encouraging the launch of start-ups. The creation and growth of high-tech enterprises can be encouraged by promoting a culture of entrepreneurship in clusters and fostering the emergence of a skills pool (managers, financial advisers, patent and intellectual property consultants, etc.) to support and work with these enterprises.

Creating public funds ring-fenced for SMEs. Another tool is the use of public procurement processes to encourage research projects in SMEs. This could involve earmarking a proportion of government funds for research in small enterprises and thus financing the launch and development of start-up technologies.

Providing financing and advice. To encourage the creation of new enterprises, specific programmes to finance enterprising projects have had positive impacts. The particular case of the Austrian LISA programme, which supports new enterprises through financing and advice, is a good illustration of this approach.

Stimulating innovation and collaboration

Fostering SME-research collaboration

Encouraging technology transfers. Closer collaboration between research bodies, universities and SMEs could generate a greater economic impact in the cluster. Start-ups and SMEs can play an important role in the development and transfer of technologies within the cluster, for example through the production and marketing of products and services which complement the technologies resulting from research. It would be worthwhile encouraging collaboration by providing structures and initiatives which encourage such exchanges.

Fostering collaboration by neutral agencies or networks. It is very important to facilitate collaboration between the players in a cluster, in particular co-operation between research organisations and enterprises (including start-ups and SMEs). Using neutral agencies or networks which serve as brokers can facilitate such dialogue.

Promoting people mobility. The creation of formal programmes to encourage mobility among professionals between industry and the academic world can be an important means of promoting knowledge-sharing and strengthening collaboration. Recurrent secondments of researchers for a given length of time to enterprises can also be a good way of familiarising researchers with the needs of industry.

Encouraging collaboration: within the cluster and between clusters

Encouraging a local innovation policy. Interactions between the players in one or more clusters in the same geographical area can be considerably reinforced by developing a local innovation policy. The importance of this approach is underlined in Medicon Valley, where public actions are designed to enhance the connection capacities between the players of the "triple helix" (government, industry, research).

Creating formal projects for collaboration between clusters. The initiative of the Oxfordshire and Cambridge clusters is a good example of the benefits which can be drawn from this approach. With the "Oxford-Cambridge Arc", the aim is to link the networks in the two clusters and weave new relations in the emerging technologies to create a stronger critical mass. The UK-Medicon Valley Challenge Programme is another good example of such projects.

Involving the non-members in the core activities of the cluster. The SMEs and very small and medium-sized enterprises (VSMEs) of more traditional sectors can often contribute to some of the cluster's core projects with their know-how. The coordinating agencies of these enterprises, such as chambers of commerce and chambers of arts and crafts, can help formalise targeted collaboration projects with enterprises outside the cluster core. A good example of this approach is illustrated by the Metis project in Grenoble.

Communicating the benefits of collaboration. It is not unimportant to use open communication mechanisms to send the message of the positive advantages of co-operation. The Vienna cluster has organised a number of "calls for proposals" in order to generate a demonstration effect by highlighting "winners" who explain the factors behind their success. The Vienna cluster also shows the need to mix a variety of policy actions to help SMEs to overcome barriers to co-operation.

Encouraging enterprise networks

Developing specifically technological enterprise networks. The inclusion of SMEs in formal networks helps to create a climate of trust between members of clusters which encourages collaboration. The creation of enterprise networks can be supported by financial contributions from government at national, regional and local level in order to unlock the development of the private sector. It is not so much the number of networks which is important as the way in which they take shape and the quality of the geographical and sectoral coverage.

Creating meeting spaces. Informal exchanges are just as important as those which take place around a working table. The provision of social meeting spaces to generate a "cafeteria" effect is worthwhile. Nevertheless, other more formal meeting spaces should also be provided such as workshops to share knowledge, and seminars and conferences to encourage dialogue.

Better marketing of products

Allowing creators to keep intellectual property. By creating policies so that "creators" in the university possess intellectual property rights over the results of research. This would encourage researchers to get involved in spin-offs or sell their ideas on the market.

Establish a marketing centre. Establishing a marketing centre for products resulting from academic research located in the very heart of university is an innovative approach recently tried in Toronto. The centre can serve as a public relations bureau by linking the economic reality of the results of academic research and vice-versa, so as better to identify a potential market.

Coordinating public polices and regional initiatives

Strengthening public-public and public-private partnerships

Creating strong partnerships. It is important to have solid partnerships comprising key enterprises, local government authorities, university institutions and the business community. That would facilitate the implementation of economic development projects and the definition of common initiatives and types of support from which the clusters could benefit.

Developing a joint strategy for the cluster. It is necessary, in partnership with key national and local players (public and private) involved in the cluster, to create a strategy for a clear vision of the way in which the cluster should evolve. This is necessary to meet priority needs, especially investment in infrastructure.

Encouraging evolution in cluster activities

Encouraging new activities. The achievement of competitive excellence in the globalised world is only attained if innovations evolve at the same pace as knowledge. It is therefore important to create policies which encourage the emergence of new activities in complementary knowledge sectors to the existing base of the cluster. The cases of Oxfordshire and Medicon Valley show how diversification is seen as a priority and is encouraged by the provision of new infrastructure, new networks and the development of new enterprise formation.

Ensuring quality human capital

Updating education and training to meet the requirements of the cluster

Adaptation of university programmes. University courses find themselves forced to evolve to meet the needs of clusters. To that end, it is essential to establish a continual dialogue between industry and universities so as better to understand those needs. Establishing a formal collaboration programme would be useful in adapting university courses in the light of identified industrial needs. The Waterloo Co-op programme is a good example.

Creating databases to anticipate skills needs. In order to improve skills levels in the clusters' new activities, it would be useful to create trend databases and produce forecasts of skills needs. This approach was successfully adopted in Medicon Valley.

Ensuring availability of talent locally

Attracting talent from abroad. Future demand for qualified workers, once identified, must be satisfied early on to avoid any slowdown in the activity of the cluster or discouraging a new player from setting up in the region for fear of a shortage of human capital. The implementation of programmes which target foreigners (as in Dunedin) or expatriates (as in Vienna) to come and work in the cluster can have positive results.

Ensuring the appeal of the area and a good quality of life

Maintaining the quality of the environment. In order to maintain the appeal of the region to attract talent, measures to improve road traffic must be taken to reduce congestion. The creation of free car parks on the outskirts of town is one of the various initiatives taken in Oxfordshire which seem to have had good results.

Facilitating access to financing

Encouraging private investment

Involving private investors in the activities of the cluster. Involving a venture capitalist as a member of a cluster governance organisation is a mechanism which can prove useful not only to link financing to the heart of the cluster but also to allow a better understanding of the market from the investor's point of view. The Minalogic cluster in Grenoble recently implemented such an initiative.

Establishing a technology transfer bureau. Supported by a policy which allows public research institutions to claim the commercial rights on inventions of their own employees, a technology transfer bureau could be established to create a bridge between innovations from the design phase and the market. Denmark has adopted a policy of this kind.

Improving access to finance. Mechanisms to encourage private financing (venture capital and business angels) need to be reinforced, both by improving the sophistication of entrepreneurs in their demand and search for finance and by increasing the supply of investor funds to clusters. In order to encourage private investors to finance innovation, it may be useful to create tax incentives specifically for funds placed in an innovation project in the cluster, with a predefined ceiling. The Wisconsin Angel Tax Credit serves as an inspiration for this approach.

Facilitating access to public funding

Improving access to financing for innovating SMEs. To increase the contribution of innovating SMEs to the development of clusters, it is important to ensure easy and rapid access to financing. To do this, it would be useful to create and publicise the existence of a "one stop shop" for SMEs. This would help in granting public funds with a minimum of delay and a minimum of administrative formalities and signposting to private sources of finance.

Creating forums to seek financing

Organising recurrent events. Isolated events do not have any long-term impact. It is therefore desirable to encourage the organisation of recurrent events which serve as platforms for investors and entrepreneurs to meet. The "Venturefest" fair in Oxfordshire and the "Forum 4i" in Grenoble are two inspirational examples.

Reducing congestion and social divisions

Addressing congestion and social inequalities resulting from the emergence of the cluster

Offering technical training programmes. The development of clusters may lead to a problem of social cohesion since it tends to increase demand for highly qualified staff without necessarily having a major impact on unskilled or low-skilled workers or the unemployed, except through general multiplier effects. Few unskilled jobs are offered while the major traditional employers are shedding jobs. Short technical training programmes can draw on the capacities of the less skilled and steer them towards new employment niches in the core of the cluster.

Investment in social housing. The growth in high-skilled jobs tends to increase housing costs and sidelines the less well paid. Greater investment in social housing could be suggested to respond to the problems of social cohesion.

Local planning policies and investment in transport infrastructure. Increased investment in local communications infrastructure and public transport may alleviate road congestion problems. Planning policies that disperse activity out of the more congested parts of the agglomeration would also help. Inspiration is provided by the efforts of Oxfordshire in these two areas.

Creating mechanisms to inform the community about the activities of the cluster

Writing and distributing newsletters to the general public. A strategy of direct and simple communication should be established in the clusters, in particular where there may be social and ethical concerns about cluster development. These newsletters to the general public must be regular and open a debate on the policies surrounding the activities of the cluster and the implications for the population.

Inform the SMEs and VSMEs about the activities of the cluster. SMEs and VSMEs outside the cluster can play a role as suppliers of goods and services to the enterprises and employees of the cluster's core. These enterprises help to boost the cluster's growth and ensure a better multiplier effect by creating jobs throughout a larger area. Better communication by the cluster governance with small firm representatives and support organisations, such as chambers of trade and chambers of commerce, could help in conveying information about needs and economic opportunities to be exploited (*e.g.* in terms of adjusting working time, products required by members of the cluster, etc.).

International models of good practice

To complement the recommendations set out above, we present a table with a series of practices and initiatives adopted to address the challenges of the clusters analysed in the book. The approaches used and the players participating in these initiatives vary from one region to another. Nevertheless, it is hoped that these practical models will serve to illustrate the recommendations in order to give a better idea of their implementation in a given context.

 CLUSTERS, INNOVATION AND ENTREPRENEURSHIP – ISBN 978-92-64-04442-5 – © OECD 2009

Table 9.1. **Synthesis of good practices identified in the clusters analysed**

Recommendation	Programme (cluster)	Approach
Encouraging entrepreneurship		
Supporting spin-outs	Best of Biotech – BOB (Vienna)	BOB is a business plan competition promoting the entrepreneurial activity of researchers in bioscience related fields. Its goal is to increase the number of young life science firms by stimulating researchers to translate their ideas into business plans.
Supporting spin-outs	Bioincubator (Medicon Valley)	The Biomedical Centre (BMC) houses this unit which aims to incubate new enterprises specialised in the areas of the cluster.
Leading a transition to the entrepreneurial university	Centre for Business, Entrepreneurship and Technology – CBET (Waterloo)	This unit co-ordinates the activities related to entrepreneurship: the education programmes, the teaching of key competences on business creation and exploitation, etc.
Supporting the launch and growth of start-ups	Wisconsin Alumni Research Foundation WARF (Madison)	This independent organisation provides logistic support on the patents, put scientists in contact with other actors and engages in the University's start-ups to facilitate their access to venture capital.
Supporting the launch and growth of start-ups	"One stop shop" LISA VR (Vienna)	LISA VR provides cluster management services to the local biotech industry. Its services include consulting, pre-seed financing, education and mediation of incubation space.
Supporting the launch and growth of start-ups	Small Business Innovation Research Programme (United States, Madison)	The SBIR funds substitute for early stage venture capital. It has a good impact on entrepreneurship in the cluster.
Stimulating innovation and collaboration		
Fostering SME-research-university collaboration	"Co-op" Program (Waterloo)	The programme offers formal exchanges of students in regional enterprises. This facilitates the technology transfers and the forecast of skills needs of the industry.
Encouraging collaboration	UK – Medicon Valley Challenge Programme (Medicon Valley)	This collaboration programme aims to link up with other global biotech "megacenters". It promotes research exchange and interaction between organisations.
Encouraging collaboration	Oxford-Cambridge Arc (Oxfordshire)	This initiative seeks to bring closer the two clusters and to build relations in the fields of new technologies in order to create a stronger critical mass in the region.
Encouraging collaboration	Metis (Grenoble)	Metis mobilises leading innovative enterprises of the textile and paper industries to participate in collaborative projects with the Minalogic cluster.
Targeting specialisation in the cluster by encouraging its development	Initiative Vinnväxt (Medicon Valley)	This initiative promotes a diversification process through the support of research in biotechnology in the food sector.
Encouraging enterprise networks	Oxfordshire Bioscience Network, DiagNox, OXIT (Oxfordshire)	These networks have benefited from the financial support of the central government and the regional development agencies
Better marketing of products	Wisconsin Alumni Research Foundation – WARF (Madison)	The WARF plays a key role in the commercialisation of new technologies in the University of Wisconsin and in the promotion of spin-offs and licences.

Table 9.1. **Synthesis of good practices identified in the clusters analysed** *(cont.)*

Recommendation	Programme (cluster)	Approach
Better marketing of products	Intellectual Property (IP) policy at the University (Waterloo)	The University allows students and researchers to keep entirely the rights of their IP. The innovators are encouraged to seek market opportunities to commercialise their idea.
Better marketing of products	The Bayh-Dole Act (États-Unis et Madison)	Universities and SMEs keep their property rights of their innovations with the support of public funds from the Federal government. This encourages the researchers to seek market opportunities.
Coordinating public polices and regional initiatives		
Strengthening public-public partnerships	AplusB Programme (Vienna)	This national programme has the task to support regions to establish centres that focus on the stimulation of new firm formation.
Strengthening public-private partnerships	Oxfordshire Economic Partnership and Oxford Trust Networks (Oxfordshire)	These two partnerships gather members of the business sector, universities, research institutes and policy makers. They carry out specific projects to support the cluster.
Ensuring quality human capital		
Updating education and training to meet the requirements of the cluster	Waterloo Plan (Waterloo)	This co-operative education programme proposes a new type of education to be offered on a co-operative basis with industry. The University shares the burden of technical training with industry and benefits from funding from the private sector.
Ensuring availability of talent locally	Database/directory of competencies (Dunedin)	A directory of competencies is made available in hard copy and on the website to demonstrate the capacity of critical mass.
Ensuring availability of talent locally	"Brain Power Austria" (Vienna)	This programme aims to attract talented Austrian scientist from abroad. The main activities include the provision of financial support, relocation services, coaching, and promotion of Austrian job opportunities.
Ensuring the appeal of the area and a good quality of life	Subsidise housing (Oxfordshire)	Housing is subsidised by the public sector for "key workers".
Ensuring the appeal of the area and a good quality of life	Grow Wisconsin Initiative (Madison)	This initiative aims to create an environment that encourages business development by creating a competitive business climate, investing and reforming regulations.
Facilitating access to financing		
Encouraging private investment	Wisconsin Angel Tax Credit (Madison)	This is a tax credit proposed by the regional government to business angels willing to invest in local enterprises.
Encouraging private investment	Minalogic (Grenoble)	The first venture capitalists became members of the Minalogic cluster.
Encouraging private investment	Oxfordshire Investment Opportunity Network – OION (Oxfordshire)	OION is a business angel network that links prospective investors with entrepreneurs seeking the early stage funding.
Encouraging private investment	Office of Corporate Relation (Madison)	This office behaves as an interpreter of and liaison between the support programmes and the businesses. It also connects new, start-up companies with the outside world.

 CLUSTERS, INNOVATION AND ENTREPRENEURSHIP – ISBN 978-92-64-04442-5 – © OECD 2009

Table 9.1. **Synthesis of good practices identified in the clusters analysed** *(cont.)*

Recommendation	Programme (cluster)	Approach
Facilitating access to public funding	"uni:venture" (Vienna)	It provides venture capital to academic spin-offs. Companies can receive up to EUR 1.1 million for a period of ten years.
Creating forums to seek financing	"Venturefest" (Oxfordshire)	This "international entrepreneurs fair" is the annual flagship event of the cluster that aims at proposing a place to meet for entrepreneurs that seek business partners or investment.
Creating forums to seek financing	"Forum 4i" (Grenoble)	This Forum provides a platform to activate venture capital for innovative enterprises by bringing together investors and entrepreneurs.
Reducing congestion and social divisions		
Addressing congestion resulting from the emergence of the cluster	"Smart Growth" (Oxfordshire)	The traffic congestion has been reduced with the establishment of the "park and ride" scheme whereby car parks are built on the outskirts of the city and served by regular bus services to the city centre.
Creating mechanisms to inform the community about the activities of the cluster	Communication on the University's activities (Oxfordshire)	To face ethical objections against the animal testing facility in the science area in Oxford University, the University introduced internal ethical review, information dissemination and legal injunctions which determine the conduct of demonstrations.

1363214R0012

Printed in Germany
by Amazon Distribution
GmbH, Leipzig